THE PHYSICS OF HEAVY LIGHT

AND OTHER LIGHT MYSTERIES

COL PARKES

ISBN: Softcover 978-1-5144-9488-2
EBook 978-1-5144-9487-5

Print information available on the last page.

Rev. date: 05/25/2016

To order additional copies of this book, contact:
Xlibris
1-800-455-039
www.xlibris.com.au
Orders@Xlibris.com.au

TABLE OF CONTENTS

PREFACE

The Preface introduces the book and its author and provides background to help the reader quickly find out what the book is about, who wrote it and why it was written. It is usually written by someone other than the author perhaps to ensure a less biased and more objective view is provided. But a good preface of a science book requires someone who understands the subject and could see some value or meaning behind some of the bold ideas proposed. It needed someone who could see the benefit of challenging existing theory and exposing new ideas to a broader public audience. But just like my first book, I could not find anyone who was readily available, affordable and as interested in this subject as I am. Perhaps I don't have the relevant and appropriate contacts or friends. So I decided to write it myself. Therefore I apologize if it seems a bit biased about the usefulness and potential benefits of this book, but here goes.

When I finished my first book about gravity, I was surprised that I had managed to get as far as I did, even if it did take a long time to complete. When I finally received a published copy it gave me a great feeling of achievement and relief. My ideas were now in print for anyone to read, or so I hoped. But so far my first book hasn't become a best seller; it simply hasn't sold many copies, not yet that is. So here I am publishing my second book. That alone is another great achievement. But this time I hope this one sells better and is read by more people, especially experts in the field. Who knows, it may also raise some interest in and sales of my first book.

I had no plans to be a technical book writer or author of any kind, so how did I get this far. Well I do have a fascination with physics in general and gravity and light in particular and I have a very questioning mind, too questioning according to many friends. I always wanted to contribute to scientific enquiry but had difficulty working out how to do it from outside so I decided to write books. Again this book is on physics but this time on the nature and behaviour of light. My first book on gravity proposed a bold new theory of gravity based on light but it was only half the story. I had to write another book to progress some unfinished ideas on the type of light involved. Light seems an equally challenging or should I say equally boring subject as gravity at least to non-physicists but you will be surprised at what I have found or proposed. The two areas of physics are very closely related, as my first book proposed and this book reconfirms.

This preface is somewhat similar to that in my first book so if you have read that, which is unlikely, you may notice some repetition. The subject of this new book is also a natural follow on from my first book called The **T**heory **Of** **L**ight **G**ravity (**TOLG**). In that book I proposed that light pressure, or shadowing, is the cause of gravity. My first book proposed a new form of light behind shadow gravity and raised many questions about properties of light and that needed more explanation. I also mentioned that my next book would be on the general subject of light. This may have created some interest for this new book. However, as very few have read my first book there is little chance of either of those two things happening. So if you want to know a bit about the author and the subject, or you just like to read a book from cover to cover, read on, it will be worth it.

MYTH BUSTING OR NOT

This book reflects my myth-busting approach or at least my questioning approach to science. It's not that I am always challenging and trying to replace existing theories just for the sake of it or just because they "seem wrong" to me. But I like to look into everything in more detail which is not easy in physics as it is a highly specialised field of science. Perhaps this approach may make me seem like a sceptic but my aim is to better understand how things work and remove mystery. In particular I don't like being told it is all too complex for you to understand. So I am always on the lookout for new and improved ideas and alternative explanations as to why and how things happen. I have a desire to simplify scientific theories and make them easier to comprehend, at least for me and perhaps others and also to improve them in the process.

Science theory and scientists need to be constantly but objectively challenged by society. Perhaps most within the scientific community accept the need for constant challenge (not always constant change) but some are opposed to it. It seems to me that some scientists, often "close the door" on alternatives too readily by rigidly adopting theoretical and mathematical complexity and denying simpler explanations are possible. This may apply to some quantum theorists who often say physics is too complex for non-experts and perhaps even for many experts to understand, but I don't share that view. We should beware of experts saying it is too difficult for outsiders to participate in meaningful analysis. Even the dumbest questions often raise the biggest challenges. Internal peer review or challenge is usually encouraged but external challenge is often just scoffed at. However, this shouldn't stop anyone from trying.

The challenge to the old earth centred universe theory was a classic example of that approach. The old model was propped up by more and more complex extensions (epicycles) to the central theory as new evidence questioned its validity. But it was eventually overthrown and replaced by a simpler, better (correct) model because of the efforts of "outsiders". Alternatives that have less complexity than existing theory (fewer epicycles as the saying goes), that explain observations and make new predictions should be encouraged. I believe nature (reality) is basically simple and follows a "lowest energy, least complexity" strategy. The physics that describes and models it should follow this rule. Hopefully these considerations lead on to physics being steered in the "correct" direction. But perhaps there is no real correct direction only a better one!

Unfortunately questioning physics from outside its "ivory towers" is not easy, especially in theoretical and highly mathematical areas such as gravity and light. The models and theory in these areas are based on extensive research and complex mathematics so just to understand the basics is a challenge. The subjects I have addressed in both books require a high level of scientific and mathematical background and although mine is limited it has been sufficient to enable me to ask some challenging questions and make some bold but realistic proposals. I have also undertaken my own limited research which was not easy. Regardless of complex quantum theory and difficult tensor maths, I have "had a go". I have tried to make an input and not just sit and complain. Time and you the reader will tell how successful my effort has been.

WHO IS THE AUTHOR?

So who is Col Parkes? Where did I come from and what is my claim to fame if any? Who am I to even review, let alone question the great work of famous physicists and even propose bold new ideas? Obviously I am not a well-known scientist or a reincarnated genius and I have never worked in a patent office. I'm surely not in the league of brilliant men such as Newton and Einstein who have developed and written great theories of physics. I am just an amateur or perhaps armchair physicist with a fascination for most aspects of physics. I have a desire to ask questions and challenge uncertainty, even in areas of physics already considered by many, especially physicists, to be too complex for "non-experts" and already well addressed.

Yes as some would have guessed, I am an Engineer by profession and a pragmatist at heart. I don't like unsolved problems or impractical or "cooked up" answers. I need to relate everything to realistic, yes even mechanical solutions. I know this will irritate the theoretical physicists, especially the quantum theorist who would say I just don't understand it at all. Just like the well-known saying about quantum theory, "if you think you understand it then you don't". Well there may be an element of truth in that because the theory may be wrong. But there is also the famous story called "The Emperor's New Clothes" about "stupid" people not being able to see things that "clever" people can see, even if they are not there or not true. I am always trying to find out if the "clothes are real" or if there are better explanations to the questions. So let me provide some more background to my interests in physics in particular and the possible reasons for my questioning (challenging) approach to almost everything.

My first real exposure to physics was through my father. I owe him a lot for my interest in all aspects of physics and technology, especially electronics. He worked for the British Navy during WW2 in radio signalling. We moved to Australia just after the war and he joined the Royal Australian Navy. He eventually left the navy and joined the Research School of Physical Sciences at The Australian National University (ANU) in Canberra. He would often take me to his "office" at the research school during school holidays. I saw and touched large, complicated and perhaps dangerous experimental equipment and met some interesting people. Perhaps it was mostly beyond my comprehension in those early days but some of it started to "rub off". I am sure such visits would not be allowed these days because things have changed (for the worse?) since then in so many ways which is sad. Bring back the hands on experience I say.

With this background, I began my lifelong love of all things technical and scientific and always wanted to know how and why things worked. I became fascinated by science, technology, big machines and especially electronics and this continues today. After secondary school I became an undergraduate physics student at A.N.U. I studied theoretical physics which was a natural direction for me to take, especially after my experiences with my father's work and my love of maths. During my studies I wouldn't take anything for granted and challenged everything and everyone, sometimes too much. I have maintained that aggressive approach of seeking clarification which has sometimes (often) got me into trouble. Unfortunately (or not?), I didn't complete my science degree at ANU and didn't become a theoretical physicist. Instead I eventually went on to become an Engineer. But who knows what might have been?

While I was at university during the 60's, pop culture, the Vietnam War peace protests and free-love movements were emerging. They were gripping many students in university campuses around the world. Students became disillusioned with society and "dropped out". I did also, stopped studying and got a labouring job to get some money, which I know wasn't really dropping out but I wanted to travel and needed to live. Then I set off to see the world and got as far as London and had a great time there. But I grew up a bit, got hungry, got a job and then got involved. Reality finally hit me so I returned to Australia to start a new life. I decided to take up my studies again but this time in engineering which then seemed a more realistic career option. It also seemed to me that engineering provided much more practical explanations of why things did what they did. The maths was still tough but the electronics was great and the outcomes seemed more tangible and beneficial. The work environment and money were also a bit better.

I graduated in Electrical Engineering with First Class Honours and joined the rapidly expanding telecommunications industry with OTC, Australia's overseas (international) telecommunications carrier). In its heyday OTC was a tremendous company and provided a great work environment. I spent most of my career in the fields of design and planning and became a respected expert in specialist fields of telecommunications. I initially resisted the pressure to move into management but eventually had to move in that direction. Everything was being driven more by marketing than technology, just like most engineering areas today.

But all good things come to an end and OTC was eventually "gobbled up" by Telstra, which was then the only national carrier. In Telstra I was lost in the noise so to speak and was eventually set adrift (made redundant) to bounce around among some of the other Telco/IT players. After a few career changes in workplace training and English language teaching I eventually left the business section and started doing my own thing. Now here I am continuing in a technical role but writing on one of the most specialist subjects in physics and even proposing bold new ideas. Is it the start of a new career or just a new interest? Where will it lead me? Who knows?

WHY LIGHT?

So why did I write this second book, especially as my first one has not yet been successful. Well again I certainly didn't write it for fortune. Any income I might make from book sales in such a specialist area is unlikely to cover the cost of publication. And again I am not seeking fame although any recognition of my new ideas would be appreciated. I am not a very good scientific writer and don't stimulate the reader much, other than perhaps getting the reader to challenge my bold new ideas to prove they are wrong. This book may not be as focused as my first book was on a particular new theory, but it does address key issues and challenge traditions. But the main reason for writing this book was to progress a proof for my proposed theory of gravity. In particular I wanted to better explain and even develop a proof for the light behind gravity. I may not have achieved that goal, but have put forward some good ideas for finding a proof

The major difficulty I faced writing this book was trying to get the science correct. Another was to make the book interesting to read while addressing complex ideas. The physics of light is complicated even for me as I am no theoretical physicist. Most physics is now based on complex mathematics and quantum theory. The problem was how to make my new ideas understandable and clearly identify how they challenge existing theories. Perhaps my approach, choice of terminology and my depth of analysis is too superficial but hopefully it is scientific enough at this early stage. It may seem too scientific for some and not scientific enough for experts. I just hope the reader will be able to understand my attempts at challenging and trying to clarify some existing theories on the physics of light. I have tried to make it interesting to read for all and hope it is not too boring for the experts. Welcome to my informal discussion style.

This book is not a text book or a reference book but is a technical book. It contains useful general information on the physics of light but only for background. It attempts to provide simpler explanations than existing theories for some strange behaviour of light. It describes things in terms of more tangible concepts without mysterious physics or complicated mathematics. While there is an element of myth busting about this book, I don't believe existing light theories are completely wrong, some are just incomplete or misinterpretations of observations. Current light theory has been good enough to help us understand what we see from the edges the universe down to almost atomic scales and many levels in between. But in my mind the explanation of what light really is and how it does what it does seems incomplete. Will this book explain everything about light, perhaps not! But as you realise the significance of my light ideas on the science of the atom and the universe you may be surprised.

(SO) WHAT?

What have I achieved by writing this book apart from some self-satisfaction? Are there or will there be any benefits from any of these new ideas on light theory? Do they solve important problems and produce practical applications? Well, unfortunately I haven't been able to find any direct benefits yet, commercial or otherwise, but I am still searching. I have thought about how I could patent any aspects of my ideas but unfortunately it seems there is nothing to patent, just like my gravity theory. But if anyone has any ideas on how benefits can be obtained and patents developed and applied for, I would be happy to work with them. Perhaps some of the ideas will again be picked up by Hollywood. They are always in search of any new material for a movie and I wish them well. Any publicity is great publicity they say.

So while these ideas may not immediately produce any immediate benefit to mankind or me, I hope they will make a valuable contribution. Perhaps some of them will be used as null experiment while others may eventually be proven wrong. But even that is an achievement, helping progress science in a way. However I strongly believe that some of my proposals will be proven correct and fundamentally change science for good, leading to beneficial developments in many areas of human endeavour. New research associated with new light could lead on to new technologies. Other savings could result from redirecting costly research into more beneficial areas. While these ideas may not directly solve the problems we now face such as the population explosion (the major one) or global warming (less significant but unfortunately more popular) they may help with the energy crisis. Let's hope they don't lead to new WMDs.

YOU THE READER

But who will read this book? More to the point, who will believe it and take it seriously and perhaps even help to progress this work. Perhaps other amateur physicists, who have a fascination with light and similar questions in their minds, will read it with interest. I also hope that it is read by experts in the scientific community. Some may look at it just for fun, others may quickly review it and dispose of it but some may take it seriously. Perhaps some specialists in light physics may be fascinated by the title and interested to find out about my ideas. Others may have the intention of disproving my ideas. Any reason would be acceptable to me. I just hope there are some who find enough value in some of my bold ideas to try and investigate them further and even help to promote some of them. To those amateurs and experts I say thanks in anticipation. I only hope you can understand my ideas. Perhaps you may even come up with some yourself and be brave enough to publish them.

SUMMARY

With this brief self-introduction of me and my second book, please read on and discover what made me so passionate about my new ideas on light. And once again, as with my first book on gravity, while I may have raised some interesting new ideas, I have not yet solved any serious problems or more specifically fully proven anything. Will any of my ideas prove successful or just become more nonstandard science theories produced by a complete unknown. Will I stay unknown or worse still, become known for the wrong reasons? Not famous but infamous! Hopefully I have set some directions that future research may take in order to formally prove or perhaps disprove my theories. This will require more expertise and resources than I am able to contribute at this stage. Is there anyone out who is as interested in these areas as I am, thinks there may be an element of truth in what I propose and can help progress this work? If so I would love to hear from you. Please feel free to send me your (constructive) comments

(email: colparkes@yahoo.com).

In the meantime read on and enjoy the complete book and turn off the light (not gravity) after you finish.

CHAPTER 1 – A LIGHT INTRODUCTION

Many times in science, when all seems quiet and scientists start to believe there may not be much more to discover in a particular field, an amazing breakthrough occurs. A new discovery is made which opens up a whole new area of science for rich exploration. Most if not all fields of science have gone through this cycle many times. Light has gone through many such leaps into new areas. Perhaps Newton gave light a new start with his research and bold ideas. Then a myriad of interference experiments provided a new direction toward wave theory. When Maxwell related light to electromagnetic fields it received a mathematical and electromagnetic redirection. And then along came Einstein who "discovered" the absolute speed of light in Special Relativity. Finally others such as DE Broglie connected it with the electron through quantum physics.

But recently the study of light seems to have "run out of steam" again. The physics of light is being studied at the periphery or in the marginal application of existing theories. There are few major headline stories about light compared with other areas of science. Has the study of light reached a wall again? Is historical light theory showing its age? Is light awaiting a new breakthrough so that it can enter a new dynamic stage of research and analysis? Does this book hold the next big idea on light? Perhaps when the significance of some of my bold new ideas on light are realised, investigated and confirmed, a new paradigm in light theory will emerge. This will then lead to the development of a new stage of light theory and new applications that will greatly benefit man-kind?

LIGHT READING

So what is this book really about? From the title you could be forgiven for thinking it may be about new business opportunities in electrical lighting, but not so. And it's not about martial arts or heavyweight lifting as the title may also suggest. Instead it is about the physics of light. This is not just another formal text book or a boring book on light theory and mathematics. But it is a technical book and is definitely intended to make a real contribution to the scientific study of light. It may not tell you all you want to know about the physics of light or even all you need to know about some aspects of light but it raises some interesting questions about some very important light phenomena. I believe it progresses our knowledge but I'll let you be the judge.

Many books have been written on the subject of light. Some provide general background while others provide more detailed scientific and mathematical analysis. They address existing light theory and range down to or should I say up to any level of scientific detail required. But few try to expose any real weaknesses in existing theory or propose any real challenges to it. Instead they seem to provide tacit support to existing theory and models and the interpretation of experiments and observations based on it. But is light theory all so complete and perfect?

I believe a more open and challenging examination of our theories on light and of the interpretation of the many observations and experiments that have been used to justify it, is required. This book sets out to do this. It is more like a series of essays on aspects of light that I find fascinating and for which I find the current scientific explanation wanting or at least in need of review. Some of the strange properties of light that have fascinated me and continue to challenge physicists even today are examined in different ways. Difficulties I have found with current standard theories are exposed and discussed. Bold new ideas are then proposed to explain some light "behaviour" in a new way. But in particular this book proposes and then examines a completely new type of light. It is a very high energy (heavy) light that is a key part of my new theory of gravity proposed in my first book called The Theory of Light Gravity.

This book is a bit different in some ways to my first book. I didn't really have a single bold new theory that I needed to explain and prove as I attempted to do in that book. And in my first book I was unable to carry out any real proof of my new theory of gravity and in particular of the new light behind it. This book attempts to progress that work by developing a better definition and understanding of it. My earlier book I also raised some questions about current light physics in areas such as red shift, wave/particle duality, photon behaviour and interference. These supposedly well understood aspects of light behaviour are discussed and the current scientific explanation for them is questioned. Perhaps I have not solved any problems just clarified some questions and even proposed many more. But that is also scientific progress, I believe.

This work is exploratory but is definitely scientifically based. Perhaps the depth of analysis is too shallow for some and too much for others but I have tried to strike a balance between the two extremes. And of course I am no theoretical physicist and am not in a position to carry out any complete theoretical analysis behind my proposals. I hope this work will follow in due course by me or others either in an attempt to prove or to disprove my theories. I am very fortunate in that I have no axe to grind, no academic career or position to protect and no demanding sponsor to comply with. On the other hand, because I am an outsider (outside the ivory towers of physics) there is little chance of getting peer reviews or expert comments through the normal channels or recognised science journals. So to help progress my ideas I wrote this book.

And while this is a scientific book, I have often used "non-scientific" language such as "perhaps", "hope" and "probably" and make suggestions based on how "I see things". But please don't let this language or style put you off especially if you have a scientific background. I also continue the trend of using more mechanical descriptions to support my ideas with some mathematical background provided where possible. But I have not produced detailed alternative mathematical explanations of any new light behaviour. There are no new quantum theory based solutions or even wave theory based ones, not yet. Hopefully this will follow in due course if not by me then perhaps by others who take up the challenge.

LIGHT PHYSICS

Inquisitiveness about light has been around for as long as man could see and reason. Questions about what it is made of, how it work and where it comes from have challenged mankind. Answers had often been proposed based on mysticism or religion. Our understanding of light, even up until quite recently was very basic with no real science or mathematics behind it. But as civilisation evolved and rational thinking took hold, especially during the Renaissance, early answers were found wanting and better ones were sought. Science evolved and found more logical and mathematical answers to questions about the real world. Sound scientific answers began to be provided about light, what it is and how it does the strange things observed.

The understanding of light was gradually developed by great physicists with new technology and a scientific approach. The physics of light was placed on a more formal footing about 400 years ago when and became a true field of science. Newton, who is perhaps more famous for his work on gravity and motion, dabbled in light. He questioned many of the informal ideas about light and placed new ones on formal footing. His views were strongly challenged at the time, especially by Hooke and the results of interference experiments. Then along came Faraday and Maxwell who showed how electromagnetism was related to light. These two fields of physics were tied together forever. Soon afterwards others were examining the new field of atomic physics, including the structure of the atom and trying to determine its relationship to light. These two fields also came together and produced many new theories of matter, energy and light. Quantum physics, which started out as quantum light, then emerged.

There is now extensive mathematical theory behind the current physics of light. But existing light theory is not "perfect" and there are many including myself who are always keen to find better explanations and improve it more. I have long held the view that light still has more mysteries than are realised. But what fundamental changes to the theory of light are possible and what implications would these have on other areas of physics. Well my theory of gravity showed one area where light plays a role never before realised. Light causes apples to fall, or at least explain why they do with equal acceleration, regardless of size or shape. Will any of my other new light ideas add to our knowledge and give light science new directions? I believe so but I will let you be the judge. Please read on. I am sure you will be as fascinated as I am with the mysteries of light I expose and hopefully be convinced of the importance of my new ideas.

Any new ideas will have to be supported by relevant analytical methods and models or at the very least come up with a formal approach for the development of the physics behind them. Of course mathematics needs to be included in any comprehensive review of physics, especially light. However, this book is an introductory one and takes more of a qualitative rather than a quantitative look at some new ideas on the physics of light. In due course more rigorous analysis must be carried out as part of a more detailed study of my ideas and I look forward to participating in this work. Contributions, thoughts or proposals (constructive please) in this regard would be appreciated. In the meantime let's take a closer look at my new light ideas.

The main areas considered in this book, or questions I attempt to answer are set out below. Key among these is of course a new type of high energy extremely short wavelength light identified in my earlier book on gravity. Then there is light bending which was the major initial proof of General Relativity, the current standard theory for gravity. Another area of light behaviour that I believe needs reviewing is the interpretation of red shift. This is a key component, if not the sole proof of the current Big Bang theory of the origin of the universe which is also associated with General Relativity and gravity. Others include the fascinating properties of light associated with media interfaces such as refraction and diffraction and the myriad of interference situations that arise and of course there is Newton's rainbow and especially dispersion reversibility.

HEAVY LIGHT

This idea naturally follows from my earlier book on gravity called the Theory Of Light Gravity (TOLG). In that book I developed a new theory of gravity or at least a whole new theory about what really causes gravitational "attraction". I proposed that it is based on light pressure, or more precisely on light pressure differences due to mutual shadowing. Shadow gravity is not a new idea but the force behind it certainly is. The force is due to light pressure from a new form of high energy, hence high "mass" and high momentum light. In my earlier book I left some or should I say many questions about this new form of light unanswered. In this book I examine this new type of heavy light in more detail, hence the book title "HEAVY LIGHT". I attempt to progress that work and answer at least some of questions I raised before about this new light.

Conventional light theory treats light as waves. But to better understand or perhaps more correctly appreciate the concept of mass and momentum, the wave/particle duality properties of light need to be considered. Light needs to be treated as a stream of particles or photons and each photon supposedly has momentum. A light photon is said to have mass which is what gives it momentum. Momentum is commonly associated with the concept of a moving mass relative to a reference frame. This is existing light theory and seems well supported by observation and experiment and causes no great dilemmas. But a particle with mass is normally considered a three dimensional entity and the concept of photon mass is complicated by a photon being a two dimensional "particle". And do all bosons or all photons have mass and hence momentum

The properties of mass and momentum and the resulting pressure of light were key ideas behind my theory of light gravity. However, gravity requires higher pressure than conventional electron light at reasonable intensities is capable of providing. Therefore I proposed that gravity must be caused by a new form of high momentum "heavier" light which I naturally called heavy light. And just like conventional light, this new light must also be electromagnetic and must also be based on some form of atomic activity. These unproven but sound scientific assumptions were used to propose light may also be caused by proton activity, not electron activity. I proposed protons may emit or capture light in a similar way to the well understood light creation process associated with electron energy level transitions. I also proposed that proton photons must have much higher energy due to the greater energy (mass) of protons compared to electrons.

Proton light is a completely new idea and is the basis of my new theory of light gravity. But what is proton light and why hasn't it been observed and identified already. There is considerable work going on in many areas of high energy particle physics, both practical and theoretical to find many types of particle, especially particles associated with gravity. Many physicists are trying to find a holy grail of physics, a grand unifying theory (GUT). This will link together all particles and forces of nature, especially gravity, in quantum and atomic physics. Finding a mass giving or gravity particle (Higgs Boson or my new light photon?) is supposedly one of the main aims. Considerable work is being carried out in this area, especially at the Large Hadron Collider (LHC) at CERN. I am following that research where possible, with great interest.

But in the meantime here is some interesting news.

"SCIENTISTS HAVE 'FOUND' THE HIGGS BOSON"!

As I was writing this book, this news headline appeared. It created considerable interest in the scientific community and perhaps some amusement for the general public. Initially it caused me some concern because I thought I have been beaten in the race to find the particle behind gravity. But it seems as though I have no reason for concern. No one has beaten me to discover any type of new boson or any type of new gravity theory. The "discovery" seems to have been more of a headline than a reality. From what I have been able to find out, the Higgs boson has not really been created and measured let alone discovered in any true sense. It is still a mathematical or theoretical animal. The headline was just media hype from an international conference on such matters triggered by some unusual observations of new collision jet streams at a critical energy level. Wow, what a close call! But the search continues, especially at the new CERN.

But what is the Higgs boson? Now I am no expert but believe it is one of the many new particles (in the particle soup or is it the particle zoo) of the latest standard atomic model. But why is it so important to physics in general and me in particular? According to some scientists it is supposedly "the particle" that "causes" gravity and is associated with "creating" mass. The proposers suggest it causes or carries the force of gravity just like a type of photon "carries" the electro-magnetic force. But is it a "conventional boson" or even a photon with all its properties? Can it really be created, captured or found at current energy levels and what impact will it have on current physics theory?" It would seriously challenge my new theory of gravity or even make it redundant unless it is the same particle as my gravity causing photon.

But this book is not about the Higgs Boson it is about light in general. What is there about light that is not already known you may ask? While most believe that the current theory of light is well founded and almost complete, every scientific theory has gaps. Current issues are mostly minor or peripheral and not considered significant enough to cause a major rethink of the fundamentals of light physics. But there are unsolved issues in all theories, especially in light theory and I believe some raise significant questions. Answering them may help progress other areas of physics at atomic or cosmic levels. So if you think existing light theory is well established and unshakeable you will be surprised. Just when you thought it was safe to go back into the light waves, something large and unexpected arises and it is white (well mostly).

SHIFTED LIGHT

What does the term light shift really mean? The phenomenon of red shift was discovered early last century. It was observed when spectral analysis of light from distant cosmic sources was compared with spectral patterns from local well known atomic sources. For example spectral lines of Hydrogen were located but found to be shifted toward the red end of the spectrum (called red shift) compared with their known position. Then cases of a shift toward the blue end of the spectrum were found and called blue shift. The general term light or spectral shift was created to cover both situations. Light spectral shift is the amount of change in wavelength of a particular source of light due to some external factors. The amount of spectral shift of a source is usually observed in specific wavelengths of light (spectral lines) but is assumed to apply equally to all light from that source. The big question was "what causes spectral shift"?

The initial theory on the cause of spectral shift was that it was due to the same wave effects as explained by a well-known theory made famous by Doppler after whom it gets its name. Conventional Doppler shift is based on the behaviour of longitudinal sound waves with a moving source or observer. Movement of a source of sound relative to a listener creates a change in the sound frequency. It is caused by compression or expansion of longitudinal waves which changes the wavelength of the sound. But is spectral shift really a type of Doppler shift related to relative movement between source and destination or do other factors come into play?

Light shift, which is due to a frequency or wavelength change, implies a change in light energy. This energy change is not associated with any change in the speed of light as light speed is supposedly absolute. Instead it is supposedly based on light (photons) somehow loosing or gaining energy (mass) due to some external influence? Doppler theory states that it is caused (solely) by differences in the (relative?) speed between the source and destination. According to Doppler theory it is only caused during light creation and capture processes, not between them. But if it happens during transit what external factors such the type of media or gravity or other fields in between source and destination are involved. Does it depend on just one factor or is it a function of many of them changing over time? And what is the implication of an expanding universe stretching space and time? Is this is the cause of all large red shifts?

SCHIZOPHRENIC LIGHT

What is the duality theory of light really all about? Is light a particle or a wave or both at the same time? Or is it just a case of both concepts applying at different times in different ways for different situations? These questions need to be addressed to determine what light is really "made of" and explain how it behaves. The current duality theory seems a bit ill-defined to me, especially the wave side of the equation. How can the various properties of light such as wave like behaviour be reconciled with the particle (photon) nature of light? Does light have a time/frequency duality along the lines of the time and frequency domains of Fourier analysis? Will this help to explain the particle/wave duality? Light as a particle seems to have some real meaning but how does it cause wave behaviour and a "light wave"? What are the real properties of a light particle called a photon. Properties such as mass and energy are at least loosely defined and linked but what about size and wavelength? How are they related?

SLOW LIGHT

The question of the speed of light is a fascinating one. Perhaps it started with the early realisation that the speed of light couldn't be infinite. It must have a finite although perhaps very high speed. Many attempts were then made to measure it. Eventually cosmic measures put an approximate but realistic figure on the speed. Then earth based mechanical means refined the figure to near its current value. Electro-magnetic (E/M) theory then derived speed from fundamental electro-magnetic parameters of the media involved. But what are these parameters and how do they change and what does that mean to the speed of light?

Special Relativity proposes that the speed of light is the same for everyone or at least all inertial frames of reference. It proposes that all space and time measurements are relative but that the speed of light is supposedly absolute. It is also known that, compared to an upper speed limit in free space, light speed may be changed (reduced) significantly depending on the media it is travelling through. There are also other ways light can be slowed down. One is said to be due to changes in space/time as per General Relativity and another due to a special state of matter such as a Bose/Einstein condensate. But how is light really slowed down in a media and can it really be stopped by a Bose-Einstein state as proposed in quantum theory?

All other types of waves use a media to travel through so it was assumed that light needed one too. But is a universal media really required to carry light? And if so what is it and what are its properties? These questions challenged physicist at the time light theory was emerging and a race was on to find out all about it. Eventually the race was lost or won by scientists who didn't believe in it. They effectively proved it didn't exist and a great scientific theory about light needing a media came unstuck. Other physicists had derived another explanation to the problem based on translations of space and time (Lorenz translations). Eventually they all came together and resulted in the light speed theory we have today. But is it complete? Are there still gaps in our understanding of the concept of light speed or light movement through a media?

BENT LIGHT

The bending of light is caused by a number of factors, mostly associated with media change. The concept of light bending at media interfaces is well known but are there others? General Relativity (G.R.) says gravity bends light. An experiment carried out during a solar eclipse was used to observe and measure the bending of star light near the sun. It was assumed that such bending was caused solely by the suns gravity as proposed by G.R. The results must have been very marginal due to the location and technology available at the time but they were used as proof of G.R. The experiment has been repeated a number of times since then with similar but not always identical results.

But are there other possible explanations for such light bending? What other media type effects may be involved that could explain such bending of light as it passes by the sun? The sun is surrounded by a thick layer of solar "gas" or plasma which has different properties to free space. Perhaps speed change due to transiting this media may have more to do with solar light bending than has been realised. The speed of light may change as it passes through this media which may cause refraction and hence bending and gravity may not be involved.

INTERFERING LIGHT

Interference which triggered the wave theory of light fascinated early light scientists as well as the general public when it was first demonstrated. But the explanation for it based only on wave characteristics, not particles, started me thinking about the possibility of other explanations for this type of behaviour. In particular the wave based explanation of the double slit experiment challenged my understanding of photon behaviour. I discuss this situation in the chapter on interference and attempt to solve this apparent dilemma. Perhaps "solve" is too strong a term as unfortunately I haven't really solved anything. But I hope I have added some insight into light behaviour and how to better understand interference from a particle point of view.

And there is the strange behaviour of light travelling between different media. The interface between two different media causes some interesting things to happen to light, including bending and dispersion. The rainbow effect is an example of this but is it reversible? And how are these behaviours related to frequency or wavelength at the various interfaces? Media differences cause differences in speed based on wavelength. These differences then cause different amounts of slowing and hence bending based on wavelength. This is readily explainable by wave theory but is another challenge for particle behaviour of light. Can I add any insight to this aspect of light behaviour using a particle approach? I believe so.

LIGHT SUMMARY

Now the scene is set for an interesting new look at light. I have identified some of the issues about light theory that have both fascinated and challenged me. I also need to progress analysis of the light that is behind my new theory of gravity. It already shows itself in terms of the force of gravity it exerts on mass but better proof of its existence and properties is required. Many other fascinating features of light that are readily explained by wave theory seem to be problematic when a particle approach is adopted. Why can't a particle approach explain all behaviours? These questions are why I wrote this book. I wanted to examine and where possible, propose different ideas on light, especially from a particle point of view.

I would like this book to be read by many people and become popular and successful, not for any financial reason, but to ensure my ideas get a good hearing. Just as before in my work on gravity, I am sure my ideas on light will raise interest but I am convinced they will also be able to stand up to scientific analysis and scrutiny and should become part of the accepted physics of light. There is a sound physical basis for them and good scientific arguments provided in support of them. I may not have succeeded in progressing the understanding of them to everyone's level of satisfaction. However, I believe I have made good progress and exposed some possible flaws in existing theory and made challenging proposals as to how things may work in different ways.

More research is required to progress my work and this will require more expertise and resources than I am able to muster and the best experts in the best environments available. I can only hope that this work is carried out soon and look forward to input and hopefully support from the wider scientific community. However as I have yet to get any feedback on my work on gravity perhaps this is optimistic. Perhaps trying to stimulate any interest in the scientific community on my light will be a challenge. So I strongly recommend that you read the whole book, I am sure you will be captivated by my ideas in one way or another. Are you willing to take up the challenge?

CHAPTER 2 – STAGES IN LIGHT

This is not a brief history of light or a story about the mysteries of light but sets out the key stages in the evolution of our current theory of the physics light. This theory has gradually evolved as our understanding of light has increased. New and improved scientific processes and technologies have been applied to research into what light is all about. This has led to great improvements in the explanation and modelling of light behaviour. But light still has many mysteries and there are still many unanswered questions. Will some of these questions be resolved or will many more be triggered by my contributions on this subject.

Fascination with light has probably been around for as long as humans could see and reason. However the understanding of light was very basic up until quite recently. Early ideas about light by Greek and other western cultures were quite primitive and not very scientific. While work on light may have been carried out by eastern cultures up to the Middle Ages few details remain. Perhaps some early eastern work and even eastern technology helped shaped the way light was thought about by western scientists in the renaissance. This may have been similar to how the helio centric model of the solar system may have emerged from early eastern work at this time.

Initially, at least in western cultures, it seems that light was considered as an emission of something from the eye. It was believed that some form of energy was sent out from the eye and bounced of an object and returned to be captured by the eye. This simple emission and reflection process was believed to be what enabled us and all other creatures with eyes, to see. This idea is not too far-fetched considering how bats "see" and how RADAR works. But what was this mysterious matter or thing that the eye emitted, what was it made of and how did the eye make it. And if the eye created this matter that allowed sight, the obvious question then became - why can't we see in the dark? The answer is simple; there is no (background) light.

Eventually it was realised that the eye doesn't create or emit anything, it only receives something from the environment. In the dark there is no light and we can't see. But in daylight there is a lot of background light directly from the sun and indirectly from reflection. External light is what gives sight, not something that comes from the eye. The eye is simply a receiver and sight needs light from external sources that bounce off objects to be captured by the eye. It was finally recognised that light must therefore be some type of substance or radiant energy that is emitted and/or reflected by objects. But what was this radiation, what were its properties, where did it come from and what was it made of? While these questions have largely been answered by modern physics they are still valid ones for consideration even today, as this book will show.

A serious analysis of light was initiated around the time of the European Renaissance. During this period, after thousands of years of darkness so to speak, a scientific theory of many physical phenomena including light started to be developed. Renaissance man "saw the light" and wanted to work out from a more scientific point of view what it really was and why it did what it did. New improved glass based optical technology such as spectacles emerged. This led to the telescope, which was initially developed for military applications. Soon afterwards microscopes, prisms and curved mirrors were developed for various uses. Scientists then started using such technology for research into light. This relationship between science and technology is an important one for both science and technology, with technology helping to progress science which then leads to improvements in technology, and so on.

Following the emergence of the helio-centric model of the solar system, which also occurred around this time, there was a rolling movement of scientific study and analysis in many fields. Basic properties of light such as reflection had been known about for ages, but the theory behind them was not well understood. The study of light was placed on a stronger footing around Newton's time. He is more famous for his work on gravity but like most physicists, also dabbled in light. He developed and used technology to investigate light behaviour and proposed many new ideas about light. But his ideas, many of which turned out to be correct, were not always well received by his fellow scientists. The basis for the theory of light we have today was established by others with little recognition of Newtons work. On the other hand his theory of gravity, fortunately, was accepted as (and was) a major revolution in scientific thought.

It is interesting to note that the development of gravity and light theories followed each other in many respects. The relationship between these two fields of physics continues today with each one often boot-strapping the other to new heights of understanding. Many famous physicists have studied and produced theories on both subjects. But surprisingly even with such a close relationship between these two theories or areas of physics, no direct connection had been identified until my gravity theory directly connected them through a new type of light.

The development of a complete scientific and mathematical theory about the properties and behaviour of light started to gather pace. Properties of light such as colour, speed and behaviour in media were under close scrutiny and were becoming more understood. Through the work of many great scientists, laws and formulae were developed to explain them. These new laws and equations and wave theory were used to predict many new types of light behaviour, most of which were quickly confirmed. The wave theory of light was becoming well established.

Then scientists discovered a connection between electromagnetism and light. An extensive new theory of light was developed based on mathematical electromagnetic field theory. This development reconfirmed the wave nature of light which had already been used to explain properties like interference. This connection between light and electromagnetism was at first theoretical and mathematical but soon became very practical. Light was being "tamed".

Finally a major step was made in connecting newly emerging atomic physics with light and this began the development of quantum theory. The particle theory of light was brought to life again and light was digitised so to speak. This was soon followed by a connection between the atom or at least the electron and light through electron orbits or energy levels. This work then led to a better understanding of atomic spectra and gave birth to spectroscopy and much more.

To provide a simple picture, here are some of key stages and great names and ideas behind the development of our current understanding of light and the physics behind it. There are no photographs but there are many other good books with pictures and diagrams of these famous people and their work. I just hope the names and descriptions are sufficient to create the images of them that I have always had. Perhaps with luck, one day my name may also be up in lights about the physics of light and gravity.

GREAT PLAYERS - THREE STAGES

1 - THE AGE OF OPTICS

The old adage of "seeing is believing" applies to this stage of the evolution of light practice and theory. Scientists were moving away from a purely philosophical approach to more practical touch and feel experimental science. They wanted to see for themselves what really happens in controlled situations. But the expression "you can't always believe what you see" (don't trust your eyes) was also true in the early days of light physics and is still relevant today. Strange light behaviour was observed but often not well understood or explained. This early science was more practical than theoretical, but theory soon followed.

THE TELESCOPE - GALILEO

Galileo didn't invent the telescope which was initially used primarily for military applications. However he developed and used it for scientific and especially astronomical observations. He saw the moon better than ever before and realised that it was not a special perfect heavenly body but just a piece of cratered rock. He also saw moons orbiting other planets, the phases of Venus and the rings of Saturn. His studies of the night sky helped to make it a popular tool for future cosmic studies. And while he did not directly contribute to the theory of light, these significant revelations contributed to the overthrow of the old Ptolemy earth centred cosmic model. Unfortunately this also resulted in the overthrow of Galileo by religious fundamentalists.

Galileo also made one of the earliest documented attempts to measure the speed of light. His methods were crude and manual but at least scientifically based. He attempted to measure the time taken for a light signal to travel between remotely located senders and observers but soon found this method impracticable. The speed of light was beyond measurement using crude timing devices, slow human response times and what were relatively short distances.

THE PRISM - NEWTON

Of course no list of scientists, even on the subject of light, would be complete without reference to the great physicists Sir Isaac Newton. While he is well-known for his work on gravity and motion, which are still the centre piece of all our basic physics in these areas, he carried out considerable ground breaking experiments and research into the properties of light and the human eye. In fact his work on light, published in a document called OPTICS, largely preceded his work on gravity but was severely criticised by colleagues including Robert Hooke. It is believed that this criticism and the self-doubt it raised caused him to delay publication of his work on light and also on gravity and forces which he published in Latin to limit readership.

Like most scientists, Newton was very interested in the unusual aspects of light that he observed in everyday life or found with the help of simple optical devices. He made his own prism and an early version of a reflecting telescope using a curved mirror. It enabled a much shorter telescope to be constructed for the same optical power as a refracting one. This was a great advance over conventional refracting or lens based telescopes which were becoming too long for practical use. This new telescope also overcame fringe colouring (rainbows) that occurs in refracting telescopes due to dispersion from the lens. Modern optical telescopes are reflecting ones, often using very large multiple mirrors, some as large as 30 metres or more in diameter.

Newton was fascinated by colour and examined how a rainbow can be made using a prism. He and was one of the first to propose the decomposition of white light produced a spectrum and that white light is made up of colours which could be separated out and recombined. His theory of colour was not very popular at that time but was eventually proven to be correct. How close he was to discovering other spectral effects such as line spectra is anyone's guess. But if he had looked at any form of artificial light with his prism, he may have seen spectral lines which may have led him to the field of spectroscopy! Another fascinating "what if"?

One of Newton's most famous theories about light was that it was not continuous but discrete or quantised. He was convinced that light was corpuscular or made of particles, although he did not seem to have any sound scientific basis for his idea. He may have been influenced by the kinetic theory of gases and the various particle theories of matter that were emerging at the time. However, this was also at a time when light interference experiments were being presented in public forums. They were probably shown at the Royal Society under Hooke as an early form of popular science. These experiments showed that light seemed to behave in the same way as water waves interfering with themselves. Light was therefore popularly and scientifically accepted as continuous and having wave like properties. The wave nature of light became accepted physics and light particles were dead in the water (waves) so to speak.

Again, while Newtons work on light may not have directly led to the development of current analytical light theory, he made considerable input to the science and was part of the revolution. It seems surprising that while he was initially very interested in the study of light, which was a very popular subject in his time, he became more interested in the study of force and motion. Why did he concentrate his efforts on the mechanical properties of nature rather than on light? Perhaps he couldn't see a future for himself in light theory. Hooke's popularity and apparently discouraging remarks about Newtons work on light may have turned him off light (no pun intended). Fortunately he went on to study and write on other great physical theories.

THE MICROSCOPE – HOOKE

Hooke made great use of the other new optical device of the time, the microscope and produced some of the great early work on microscopic biology. He drew some incredible images of micro life from observations using his primitive microscope and published them in a book called Micro-Graphia. This work had a major impact on the use of the microscope for studies of nature from that time on and helped to make this field of optical science popular and important.

INTERFERENCE - HUYGENS, FRESNEL, YOUNG, etc.

Many experiments were being conducted into the phenomenon of light interference using optical devices. Names such as Fresnel with his lenses and Huygens with his mirrors come to mind. However, Young's double slit experiment is especially famous. This experiment led to the almost complete acceptance of wave theory as the only explanation of all light behaviour. A growing body of mathematical wave theory was developed from this work. Young's experiment has also been carried out using even more conventional particles such as electrons, which also have wave like properties according to quantum theory. They all produce the same types of interference results. But questions still remain about what is really happening.

2 - THE ELECTROMAGNETIC PHASE

The next surge in activity that was important to the evolution of the physics of light was the connection with electromagnetism. It was initiated by Faraday and others and then completed theoretically by Maxwell with his famous equations. This theoretical work was followed by a string of practical discoveries and a whole new era of electromagnetic radiation technology emerged. Light was "brought under control" which led to many commercial applications. Today electromagnetic (radio/light) technology has many uses, especially in telecommunications which has greatly benefited civilisation. We know how the rest of the world live in real time and they are very like us so we don't need to fear them anymore and have a new world war.

ELECTROMAGNETISM AND LIGHT - MAXWELL

James Clerk Maxwell was one of the greatest physicists of the nineteenth century and of all time. He built on the practical work of others such as Michael Faraday who had developed the basic concepts of field theory. However, Faradays lack of mathematical skills held him back while Maxwell used his great knowledge of mathematics to develop the theory of electromagnetism that we have today. He combined the separate E/M field theories into a unified E/M theory and derived a relationship between light and electromagnetism. Maxwell's famous equations "proved" E/M waves exist which set the electromagnetic theory of physics alight and on a new path. He also studied coloured components of light and supposedly made the first colour photograph of a piece of coloured material, tartan of course.

Maxwell's equations express; how electric charge produces electric fields; how changing electric fields lead to magnetic fields; how magnetic fields created by electric currents exist in dipole form without magnetic monopoles and how changing magnetic fields produce electric fields. Some of the concepts and equations may have been around before Maxwell but he uniquely re-derived them mathematically using Faraday's field concepts. With help from other physicists and mathematicians such as Heaviside, the equations were reduced to the famous four and put into the notation used today. They are well known to physics and electrical engineering students.

MAXWELL'S EQUATIONS.

$$\mathbf{DIV} \quad \mathbf{E} = \rho/\mathbf{e}_0$$

(Electric charge density ρ creates an electric field E (radial?))

$$\mathbf{DIV} \quad \mathbf{B} = \mathbf{0}$$

(Magnetic fields B (closed) exist, not magnetic mono poles)

$$\mathbf{CURL} \quad \mathbf{E} = -\delta B / dT$$

(Changing electric field creates moving magnetic field)

$$\mathbf{CURL} \quad \mathbf{B} = \mathbf{m_0(J} + e_0\ \delta E / dT)$$

(Changing magnetic field creates moving electric current)

When these basic equations were solved they produced an electromagnetic wave solution which had a surprising result. The speed of waves was found to be related to basic properties of the media through which the waves propagate the magnetic permeability and electric permittivity. The fact that speed was a simple function of these two natural properties was surprising but of greater surprise was the speed itself. When the speed was derived it was found to be the same as the existing experimentally measured speed of light. This amazing discovery was perhaps one of the most important aspects of his work and made Maxwell and others realise that light is a form of electromagnetic (wave) energy.

Maxwell's equations "proved" light was electromagnetic and further supported the theory that light must be a wave, an E/M wave at that. But it is interesting that this theory provided no indication of quantisation of light or the existence of photons. It seemed certain therefore that light could not be a particle. In fact Maxwell's equations and the wavelike properties derived from them were assumed at the time to be the final death knell of quantised light, but it rose again like the Phoenix.

But can Maxwell's equations be associated with quantum theory? Can light quantisation and the properties of a photon be deduced from a quantised version of the electromagnetic theory? Is there some as yet unknown quantising field factor that needs to be added to the equations to arrive at this outcome? This is an extremely interesting topic and one that I have wrestled with since I learned of Maxwell's equations, light quantisation and photons. I am sure it has been addressed and I am currently reviewing this area with great interest. Perhaps a new understanding may be obtained through my proposals on light

THE ETHER - A LIGHT MEDIUM

One outcome of the new wave theory of light that challenged early scientist was the question of how light waves travel. At that time it was "sound science" to believe that waves required a media to travel through. But what was the media that carried light waves? Eventually a new media called the Aether (luminiferous ether or just ether) was "created" for light waves. It supposedly permeated through all space and matter and was permanently in and around us. Many top physicists of the day supported the idea. However, support by many experts doesn't mean proof (still an important message even today).

Extensive research was conducted to find and measure the properties of this media. It was a key scientific question of the day. But after much effort no one succeeded in detecting, capturing or measuring any of its effects. It was then suggested our orbit around the sun should lead to movement of the earth through the ether. The earth moved at a considerable speed around the sun and this speed was reversed as the earth moved in opposite directions every six months. It was also assumed that the ether would surely be fixed or "stationary" so such a large change in the speed of the earth could cause a detectable change in the measured speed of light. This triggered a search to detect any speed changes using a light phase change detector called an interferometer.

THE ETHER EXPERIMENT - MICHELSON/MORELY (MM)

Michelson and Morley, two great American physicists, constructed a large interferometer on a stabilised, rotatable platform. This enabled interference patterns to be created between light that had travelled along different orthogonal (right angled) paths in different directions at any time. The device was built to be sensitive enough to detect the very small phase differences that may be caused by a change in the earths speed. Perhaps they were looking for the wrong effect.

Of course, the results are well known; there was no change in the interference pattern at different earth orbital positions over years of observations, regardless of how the machine was rotated or orientated. The results were immediately interpreted by some to prove that the ether didn't exist. However, other physicists still supported it and developed explanations for the results. Interestingly the results were effectively interpreted in different ways by two famous physicists Lorenz and Einstein although Einstein supposedly said he wasn't aware of this result.

Lorenz said the ether still existed and the speed of light varied but this couldn't be detected because movement through it caused contraction of the measuring devices. This contraction then automatically compensated for the speed change. He developed a theory of length contraction (Lorenz contraction) to explain this effect. On the other hand Einstein said light always travelled at the same speed in any direction for any observer. He "proved" that light didn't need a media and applied Lorenz contractions in a new way in his Special Relativity theory.

The MM experiment has been duplicated many times since with modern equipment. The results are always the same. No interference pattern change has ever been detected over time and space. Today lasers are used with very long paths and very accurate electronic detectors, not just to detect speed change due to some form of ether but for other reasons, especially to try and detect such things as "gravity waves", whatever they are. Until very recently nothing had been detected except earthquakes and heavy traffic. However a news report just in said "gravity waves" had at last been "detected". But perhaps this is just like the Higgs boson "discovery" and is just over enthusiastic misinterpretation of some unusual observations. Time and my ideas will tell.

But light theory then started generating new waves itself, in a more practical sense.

RADIO WAVES - HERTZ, MARCONI

Perhaps one of the most famous scientists who took up the challenge to apply the results of the theory of E/M waves was Hertz. He experimentally created radiation or light waves using a simple spark gap generator. He confirmed that controlled transmission and reception of E/M waves, at least at lower frequencies, was possible. This technology was further extended and applied by Marconi for practical uses. He developed many commercial applications early last century, the first being a trans-Atlantic radio communications service. This then led to a whole new field of telecommunications technology, the area that I eventually entered.

SPECIAL RELATIVITY - EINSTEIN

Einstein is famous for his theory on the speed of light and its effects on time and measurement or vice versa. It seems he set out to study how Maxwell's equations relate to moving observers and fields and what these may do to the speed of light. He supposedly conducted a thought experiment by "riding a light wave" (an early form of light surfing). His publication called "On the Electrodynamics of Moving Bodies", which became known as Special Relativity (S.R.), shook both the scientific and non-scientific worlds with its profound outcomes.

He arrived at the startling conclusion that the speed of light is the same for any observer travelling at any (sub light) speed in any direction. That is, light speed is invariant for all inertial frames of reference. Some say that he started with this postulate as a-priori "knowledge". In a way Einstein confirmed that the speed of light is absolute, not relative. Somehow this seems contrary to the whole idea of relativity where physical properties such as time and distance (space) are relative to the frame of reference they are measured in. However it is now universally accepted as the correct interpretation of light behaviour. But does it apply to all types of light?

Einstein's theory agreed with Lorenz that measuring devices in moving frames of reference contracted, but not because of ether pressure but because of the constancy (absoluteness) of the speed of light. This interpretation by Einstein became the key to his Special Relativity. It greatly challenged our understanding of light once again and eventually killed the ether theory or the need for a light carrying media, for good. But it raised many more questions about the relationship between light and the emerging field of atomic physics.

3 - THE ATOMIC AGE

The third stage in the evolution of light theory is linked to the atomic age. It could be said this stage began with a bounce not a bang. The atom is too small to directly observe measure and analyse in a physical way but the peculiar outcomes of experiments were noted. The atom came under attack by various radioactive particles and unusual trajectories (bounces) were observed. These experiments shed new light on the structure and components of the atom and on light itself resulting in a new atomic model.

Mathematical modelling finally connected light with the atom. From this theory many new atomic discoveries were made and a new atomic based model of light or is it a light based model of the atom, was developed. It had a profound impact on the theory of light and atomic physics and gave birth to quantum theory. Many were involved in the physics of the atom and light but here a few of the key events and players.

THE ELECTRON AND "PLUM PUDDING" ATOM - THOMPSON

The first atomic model was proposed by J.J. Thompson after he discovered the electron. It was a simple "ball of matter" with the electrons scattered throughout like plums in an English plum pudding. The name stuck (like a plum pudding does sometimes) and it became the accepted model around the start of the 20th century. The plum pudding atom had to contain other as yet undetermined matter for charge balance to make it neutral and to provide the necessary mass. But this model couldn't properly explain some unusual experimental observations.

Other scientific discoveries contributed further to the development of the atomic model, including the discovery of radioactivity which Thompson was also connected with. It was noted that some substances produced radiation and changed into new elements. Three types of radiation "particle" were identified: a heavy positively charged one called an Alpha particle, later found to be Helium nuclei; a lighter negatively charged one called Beta radiation, later identified as electrons; and high energy Gamma radiation (high energy photons?). These became the physicist's new weapons of choice in attacking the atom to discover its secrets.

ATOMIC NUCLEUS - RUTHERFORD

Ernest Rutherford, a New Zealand physicist working at Cambridge University (Newtons old stamping ground), was very interested in radioactivity. He discovered the concept of half-life for radioactive elements which went on to be of great use for geological aging but bad news for some types of nuclear pollution. Rutherford was fascinated by radioactive decay and atomic experiments that created new elements. He liked to "shoot" radioactive rays or particles at matter, especially gold foil targets, to see what would happen. Perhaps he was trying to create other elements such as lead from gold, in reverse of the alchemist dream of past ages.

To his surprise, when alpha particles were fired at thin gold foil, very occasionally large deflections were observed. This lead to the now famous expression that went something like, "it was like bouncing a cannon ball off tissue paper". Rutherford's results seemed inconsistent with the plum pudding atom which was considered as dense as "tissue paper". He proposed that such high angle Alpha scattering could only mean the atom must have a concentrated and positively charged core. A ratio of core size to atom size of around several thousand was derived indicating how many deflections had to be observed. Who said physics experiments were easy!

This experimental result killed (ate) the plum pudding model and produced a new atomic model with a very small central nucleus. The atom was believed to be surrounded by a lot of free space but it still somehow contained electrons moving throughout this space. This model had also been proposed by Japanese physicist Nagaoko around the same time. Discussion about this new model and its dynamics became a major topic among atomic physicists. Questions were raised about how electrons "behaved" and how the core or nucleus may be structured. Physicists started searching for the components of the nucleus and wanted to know how it all held together.

Soon after the solid central nucleus was proposed, a new subatomic particle was "created" to balance electron charge as the atom was known to be electrically neutral. The real discovery again came from Rutherford scattering, this time by firing alpha particles at nitrogen. This generated positive particles that behaved like hydrogen nuclei. They were found to have an equivalent charge to the electron in terms of magnitude but with opposite polarity. The new particle, eventually called a proton, was also "given" larger mass to make up the overall atomic mass of elements. Protons were proposed as being contained in a "central solid" nucleus.

The neutron was the last of the standard atomic components to be discovered. Something was required to make up for the mass of all atoms larger than Hydrogen and Rutherford proposed a neutral particle later called a neutron. But its discovery was delayed because it could not be observed in existing atomic scattering experiments, could not be controlled or deflected by E/M fields as it was electrically neutral and was easily stopped, even by vapour within a gas chamber. New methods of detection or capture were required to enable the neutron to be "seen" so that a better picture of the atom could be developed. Technology again came to the rescue.

PLANETARY ATOMIC MODEL - NIELS BOHR

The now well-known "planetary model" of the atom with a core nucleus of protons and neutrons together with orbiting electrons was initially developed by Bohr in the early part of last century. This planetary model, shown in the diagram below, then became the standard macro atomic model. It provided a sufficiently accurate picture of an atom at a basic physics level and was used as a basis for further work on atomic structure. This model is still taught in schools and is commonly used by the popular media to represent an atom.

THE "PLANETARY" ATOM

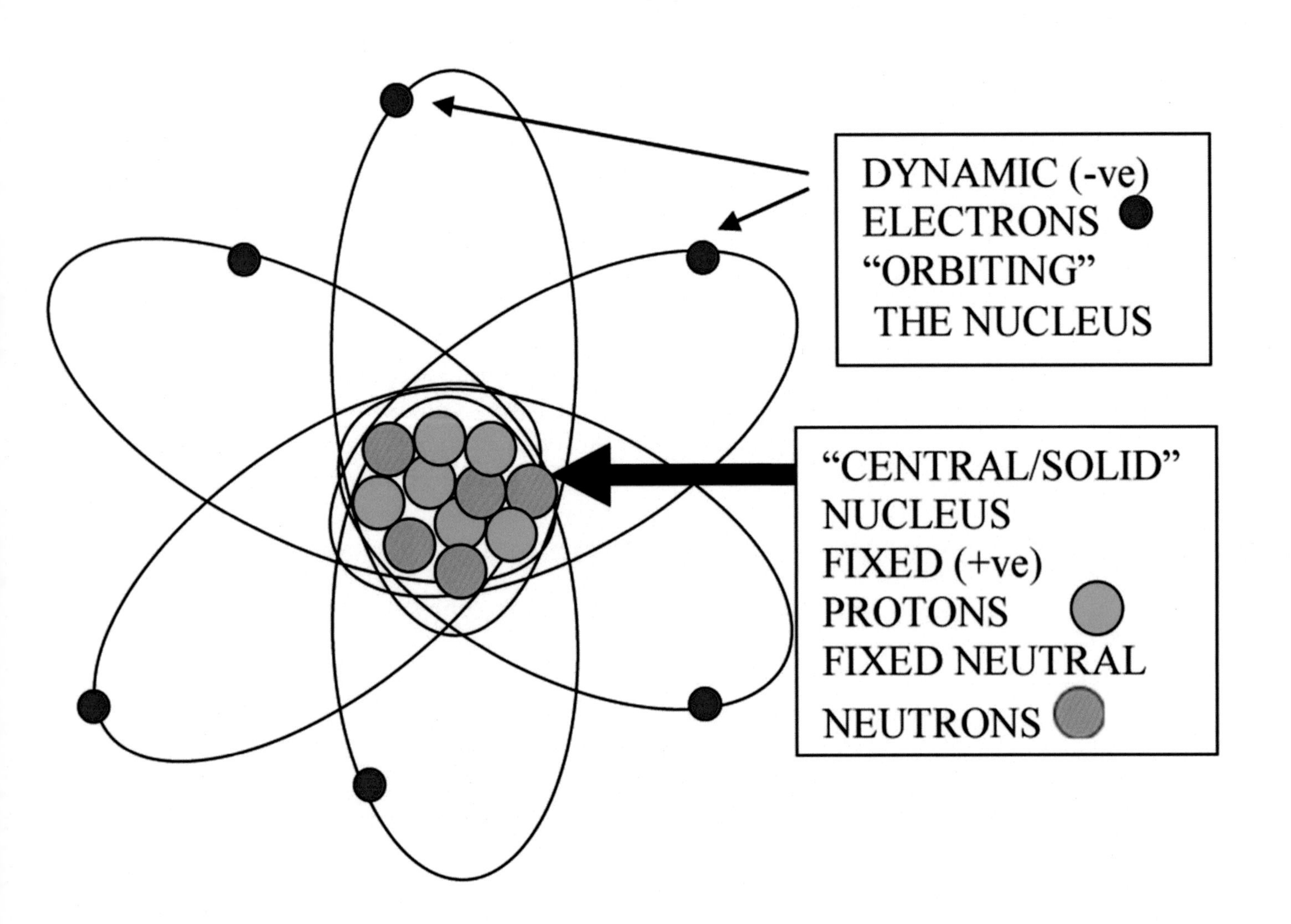

THE OLD "PLANETARY" ATOMIC MODEL
(NOT TO SCALE)

Unfortunately, this elegant simple model was soon found wanting. It couldn't explain many experimental results and observations and conflicted with existing theories. The "orbiting" electrons had stability and energy problems and no suitable explanation of how the electrons stayed in orbit without acquiring or releasing energy could be found. In addition the new photoelectric process of photon creation or capture could not be satisfactorily explained by this simple model. The problem was that circulating electrons should create current and associated fields and perhaps create energy emissions and light. But this was not happening or was it?

Many physicists then became actively engaged in the race to better understand the atom and improve the model. The component particles of the atom, their dynamics and the forces involved became the big atomic questions of last century. The problem was that the atom and its components are so small and so difficult to work with that a new type of physics was required. Existing particle firing techniques were inadequate as they were too low in energy and not controllable. More specialised technology and machines were developed to "see inside" the atom by smashing it and observing and measuring the outcomes. But unlike Humpty Dumpty, atom mashing not only breaks the atom but helps to put back together again, at least in theory

To meet these requirements a whole new "industry" of atom smashing technology was created which is the mainstay of atomic physics today. It started with a small and simple electron accelerator called a cyclotron, the first model of which was almost hand held. Then it expanded from there up to massive machines the size of a small country such as the Large Hadron Collider at CERN near Geneva Switzerland. These facilities are the focus of most work on atomic components, structure and forces and an ever increasing range of sub atomic particles. They employ many physicists, use many resources and consume a large part of the global physics research budget, all of which needs regular justification (read – "new discoveries").

A number of famous physicists carried out research into the atom, especially energy concepts, electron activity and light but I have singled out a few who played fundamental roles.

LIGHT QUANTISATION - MAX PLANK

Max Plank, a famous physicist at the turn of the 20th century, was interested in the energy aspects of physics, especially black bodies. He set out to understand the so called ultra violet (UV) catastrophe. Existing wave based light theories created an uncontrollable runaway energy problem for higher and higher frequencies of light, especially UV light. Plank solved it by proposing that the energy of light must be related to frequency or wavelength, the higher the frequency the greater the light energy and that light must also be quantised in discrete packets or bundles of energy. Planks famous equation directly relates the parameters of energy and wavelength for each "bundle of light" with a constant called Planks constant, as set out below;

$$E = h * \upsilon$$

Where E is the bundle of light energy, υ is frequency and h is Planks constant.

This equation together with the assumption that light was quantised was sensational. Supposedly even Plank didn't believe it was correct. His new theory solved many outstanding problems and was subsequently "proven" to be correct. It went on to become a cornerstone of light physics but it also created a new problem. What was this new quantised light particle and how did it relate to existing wave based light theory?

PHOTOELECTRIC EFFECT - EINSTEIN

While Einstein is known for his work on relativity he also studied light. Any review of light would not be complete without reference to his work in the area. He was a friend of Plank and was also very interested in the properties and energy of light. In one of his famous 1905 papers he put light or the photoelectric effect under the microscope so to speak. He proposed that the only scientific explanation for the strange photo-electric experimental results being observed was that light was corpuscular or quantised in "packets" later called photons. He also proposed that Planks work could be extended to derive the energy of each packet (photon) according to its frequency. Quantum theory was born (not Max Borne, he came later). Einstein was eventually awarded the Nobel Prize in physics for his light work, not for his heavy relativity.

Plank and Einstein were instrumental in resurrecting the particle (quantum) theory of light. Now quantum theory is a mainstay of modern physics. It could be said that light physics was digitised by quantum theory. It has been used to explain many physical phenomena and to discover new ones and will continue to do so. However it was recognised that light somehow still had continuous wave like properties. This re-established the dilemma between the continuous wave nature and the discrete quantised nature of light. The reconciliation of these two properties was a concern to physicist at the time and still challenges physics students and even experts today. I wonder what impact my new ideas on light will have on this age old dilemma.

About this time quantum theory was being developed and wave based mathematical ideas were being used to further explain atomic structure. The quantum mathematical model of the atom slowly took shape and electron energy levels and orbits were examined using these new ideas.

LIGHT AND ELECTRON "ORBITS" - DE BROGLIE, OTHERS

The idea of using wavelengths to determine orbits was a major breakthrough. This amazing proposal was made by French physicist Louis de Broglie. He believed that all particles, not just photons, but also electrons and all other matter have dual characteristics, that is, wave like or particle like. He proposed that electron orbits (circumferences?) should be harmonically related to the energy and hence wavelength like a standing wave in a string. He then derived the wavelength of an electron from Planks equation using its mass and energy and hence derived rules for determining electron behaviour. These were found to closely match existing predictions of electron "orbital" behaviour and his proposal became an instant success.

But perhaps the most surprising outcome was the connection with spectral analysis. Suddenly many of the lines observed from the sun or local controlled sources showed a pattern that aligned with the energy levels proposed by de Broglie. Using Planks equation the known wavelengths of the lines and the energy levels and "jumps" for the electrons in the atoms of Hydrogen and Helium were related. The whole concept of light and electron activity and the spectrum were being connected as never before. The planetary model of the atom was quantised.

Other atomic discoveries helped further the understanding of the physics of light. The range of electron "orbits" was formalised for atoms and a whole new table of electron positions and energies was established. The allowable orbits, energy levels or orbital jumps of an electron in an atom were related to photon energy and spectral lines. This important discovery led to many new aspects of atomic, spectral and quantum theory. The atomic model became very mathematical and all particles are described by a range of quantum numbers or properties. Of course this includes the photon which is a special type of particle called a boson.

In the latest quantum model (QED?) the photon is now a gauge boson defined by a set of quantum numbers including a spin of 1, zero mass (zero quantum theory mass but still with momentum), zero charge and some others a bit beyond the scope of this book. The exclusion principal and also possibly the uncertainty principals also do not apply to photons. The current QED model of an electron is based on extensive theoretical work and experimental results but it is very complex. There seem to be new developments which frequently change some of the fundamental quantum properties of these two particles. Perhaps my new ideas could have a significant impact on thinking and research in this area.

Many physicists are trying to relate all atomic (quantum) particles and cosmic forces, fields and theories together into a unifying theory. Most work is primarily theoretical and is trying to analyse unusual outcomes of high energy atomic experiments. String theory, developed as an elegant multi-dimensional mathematical model in an attempt to produce a unified theory, predicted multiple universes and other meta-physical wonders. It was going to solve everything but progress has not been very spectacular. The Higgs boson was also "created" to answer unknowns but it also seems to have produced limited results. Apart from these elaborate theories, little progress has been made, especially on how light and gravity fit together. Now I agree that I don't understand theoretical physics well enough to be able to constructively comment, but it seems that no recent work on light has been worthy of inclusion in any Book of World Records. No new major alternate theories or discoveries have been made. Is this about to change?

4 A NEW AGE? – "HEAVY LIGHT"

Is there to be a new age in the development of light theory? Will my proposal of proton photons and light gravity change the direction of light research and lead to new discoveries? Will there be new cosmic and atomic theories based on some of the ideas discussed in this book? Will this work set a new course in fundamental physics called the age of heavy light? And how do some of my ideas relate to other major physics mysteries such as Dark Matter and Dark Energy. Are these just epicycles in an incomplete (or is it incorrect) Big Bang cosmic theory or will they be upgraded by my idea on new light or even be replaced by it? Watch this space, perhaps one day with luck the name Parkes may also be in light physics.

CONCLUSIONS

This chapter discusses three stages of development in our understanding of light and the theories behind it. But the major developments are past history. Following quantum theory there seems to have been little if any successful progress in the development of new light theory. Work is now concentrated in trying to find new atomic physics particles, including new bosons and hence perhaps new photons. Big atom smashing machines are throwing faster more energetic particles at each other in an attempt to progress physics but there have been few if any major verifiable outcomes. Physicists are busy trying to connect forces, fields and particles together under a simple, complete so called Grand Unifying Theory. But this work seems to have faltered and little real progress has been made. String theory was a possible candidate but progress has not been as much as the initial hype suggested. The Higgs boson and Gravity Waves are, I believe, also challenged discoveries that may be just "timely justification" for continued costly experimental research.

There have been some developments into the application of existing light theory which have helped progress areas of science and technology. Lasers, liquid crystals, light emitting diodes (LEDs) and CCDs in cameras are some that come to mind. Many of these applications are also following a digital path. But mostly only small steps have been made in a field few major developments. There seems to have been little if any progress in the development of improved light theories or new interpretations of known light behaviour. Are current light theories correct and complete? Have all questions been answered? Is current research seeking answers to the correct questions? I believe the answer to these is mostly NO!

So what is the future for light physics? Is it going to be bright? The development of light theory occurred in steps often based on unrelated discoveries. Will this approach provide a way forward for future light theory development? Will there be a new phase in the saga of light that will reinvigorate the development and application of light theory? If so what will be the basis for this next phase? What new discoveries will trigger the next leap forward and lead to major changes to existing theory and even the emergence of new theories? Perhaps my new idea about heavy light will open up new fields of light physics and trigger new work that will lead to many new discoveries and useful applications. Read on and find out more about some of my bold ideas and perhaps be part of this fourth stage light physics revolution.

CHAPTER 3 - LIGHT MEASUREMENTS

"And then there was light" to quote or misquote a well-known saying on the origins of things. But what is light and where does it come from? More specifically how is light created, how is it destroyed (captured?) and how does it "behave" or do what it does in between these two events. These are some of the questions about light that physicists have addressed in a scientific way. As a result, our understanding of light or more generally of electromagnetic radiation is now well-established. The current scientific theory of light explains these things in some detail, but can it be improved? I believe there is always room for improvement, even in light theory.

In scientific terms, light is "electromagnetic waves" and electromagnetic waves could also be called light. The general term "light" is used throughout this book to represent all types or perhaps more correctly all frequencies and all sources of electromagnetic (E/M) radiation. Of course only a limited bandwidth of light is visible to the human eye. But the bandwidth of light as we know it extends over many orders of magnitude from very large kilometric wavelengths down to wavelengths of almost atomic dimensions. But it can still all be called light as the same basic properties apply for all wavelengths. Light supposedly "behaves" in a similar way no matter the frequency or wavelength (colour). However, wavelength often plays a role in determining the degree or amount of some types of "behaviour", especially in media.

The basic properties of optical light that we can readily observe such as transmission, reflection, refraction and absorption are reasonably well understood. These observed phenomena of optical light are also known to be representative of the general behaviour of all forms and all wavelengths of light. Therefore what is discovered or observed to occur for one form of light is generally taken to apply to all forms of light across the entire E/M spectrum. Slight variations in behaviour may arise as a function of wavelength due to the properties of the media involved. This behaviour can be described by the wave like behaviour as well as the quantised particle like behaviour. But is there more to light than the relatively simple physics of waves or particles? Quantum physics has now taken over! Light is now in the hands of chance, or is it?

While there is extensive quantum mathematics behind the physics of light, this review of its properties and the ensuing discussion of my new ideas are primarily qualitative, not quantitative and not completely "quantum-tative". There are no references to boson quantum numbers or properties such as spin, but some properties are discussed using a somewhat mechanical approach. In due course, so alternate ideas or theories can be properly accepted, they may need to be fully analysed using relevant formal quantum methods and models. But at this stage my new ideas are presented and analysed using a simple set of physical properties or comparative metrics and ideas on possible photon behaviour. This analysis should be sound enough to raise interest in some of the areas I have addressed and in sufficient detail to challenge current light theory. That will have to do for now.

KEY PAREMETERS OF LIGHT

Observation and measurement of defined parameters is central to evaluating and proving any scientific theory. So to enable existing light theories to be discussed and new ideas about light to be raised a set of measureable properties and behaviour need to be identified and defined. This chapter builds on the history and development of the physics of light from the previous chapter to identify a set of properties as a basis for discussing and analysing alternate light theories. These properties include most of the well-known ones used in existing light theory.

Perhaps one of most fascinating properties of light is its finite but very high speed. According to Special Relativity, this speed is the same for all observers and is supposedly the maximum speed at which information (energy) can travel. But there are many questions about light speed and what may change it. Then there is the question about the origin of light, where does it come from, what makes it, what takes it away? Another vexing issue is whether light is a particle or a wave or both at the same time? What are the implications of this dual nature of light? How do particles behave like waves or vice versa and how do particles or waves interact with each other and the media? And how does light, or a light particle, have mass and momentum and create pressure? Finally what creates a rainbow, what makes up the spectrum and how continuous is it? How wide are spectral lines from specific atomic activity and why do they shift sometimes?

These are only some of the key properties and questions about light that have been bouncing around for a long time. While many are important in the measurement of light behaviour a few are worth special attention due to their relevance to my new light theories. These are;

- ***ORIGIN - light creation and capture processes,***
- ***DUALITY - wave/particle (photon) reactions,***
- ***SPEED - finite (constant?) very high speed,***
- ***MASS – energy, momentum and pressure,***
- ***SPECTRUM - spectral effects and red shift.***

ORIGIN - WHAT MAKES LIGHT

There are many sources of light or radiation in the cosmos. But on earth the sun is our most important source of light. While sun light is natural, there are many light sources on earth which are artificial or man made. This leads to the important questions of how is light created? To answer this question physicists eventually came up with a theory about light creation linked to the atom. Atoms are part of everything we see but they play a critical role in creating the light that lets us see. They also create "light" at other non-visible wavelengths that lets us do so much with technology. But to understand how this happens requires a closer look at what makes an atom, how it is constructed and how it behaves dynamically.

Light all happens because of the dynamics of the atom or at least some of its components. It was found that electron activity was a key component of light creation with electrons changing energy levels to capture or create light. This process was observed to be quantised into specific energy amounts. Each quantity of electron energy was proposed as creating a quantum of light energy with a specific wavelength or colour. This led to the atomic model of "orbiting" electrons in quantised energy levels and other aspects such as exclusion. So let's examine this atomic model, its dynamics and the link with light in more detail.

ATOMIC MODEL AND LIGHT

Physics experiments, observations and logical interpretations started to describe and define a model of the atom. The atom is of course too small to see but its behaviour can be influenced, observed and measured which is key to obtaining a better scientific understanding. At the start of last century the first model was a small amorphous ball of "soft centred" unknown matter with electrons somehow scattered throughout. This was quickly replaced by a "hard centred" model following Rutherford's scattering experiments. This then became the physical "planetary" model which is still the basic physical atomic model used today. This has now been replaced by a quantum mathematical model which is beyond the scope of this book. However the physical planetary model is still basically useful for many discussions about light and atomic physics.

By mid last century the theory of the atomic model with a central solid nucleus was reasonably well developed. It proposed that the atom consisted of three subatomic particles, the electron, the proton and the neutron in defined positions and with various forces holding them all together. The electrons were considered to be dynamic and in "large orbits". The other two main subatomic particles the proton and neutron were believed to be "static" and contained in a solid central nucleus. A set of four forces (now three?) were defined to hold the particles together. The nucleus was considered static and stable, apart from "rare and random" nuclear decay observed from some larger atoms. This breakdown caused radioactivity and created other elements.

Of the original three sub-atomic particles, only the electron was considered to be stable and indivisible (perhaps apart from special quantum effects?). At one stage it was suggested that protons may be capable of spontaneous decomposition into other particles. However extensive analysis challenged this and the proton and neutron are now considered very stable. Then higher energy atom smashing machines produced some strange "events" which indicated the existence of other unknown particles. Every physicist on the block wanted to discover a new particle and this created a particle zoo. However, perhaps many of these new particles are more imagination (theory?) than reality. Most of these new particles were not as stable as their parents and had very short lifetimes and were very difficult to create and detect let alone measure. And it was assumed that most of these new particles came from the proton or neutron as the electron was and is still considered indestructible and there were no other obvious sources.

From these results and associated new mathematical modelling based on quantum physics, it was proposed that the two nuclear particles, the proton and the neutron, may not be "solid" or fundamental after all. The current standard model proposes that protons and neutrons are made up of even smaller components called quarks. These new "particles" were given interesting names like charm, strangeness and up/down ness and were allocated quantum properties or numbers including fractional charge! New mathematical models have been produced for how these new particles combine to form protons and neutrons. This area of physics is under intense analysis and review especially with experiments using even higher energy particle colliders. But perhaps the proton and the neutron are not subdivisible after all and quark theory is just another way of explaining some unusual observations (just like Ptolemy epicycles?).

It always seemed likely to me that the neutron is just a "complex" combination of a proton and an electron with some form of "binding energy" to allow for the "measured" very small mass difference. The question then becomes do they fully combine or do they retain their original structure and only combine in a very close coupling? But if this model is correct, why has a neutron not yet been broken up into an electron and a proton in high energy collisions. Perhaps it is for similar reasons why the neutron was difficult to discover in the first place, it is neutral, hard to accelerate and collide with? And why hasn't a neutron creation process occurred or been observed from electron-proton collisions, or has it? And while quark theory may be "accepted" theory for protons, a simple stable, solid, "indivisible" proton, just like an electron, is the particle model I have used as the basis for new light. This theory may need to be reconsidered in relation to the "standard" model of a proton with quarks in due course but that is for further study.

During the early development of the planetary atomic model a whole new field of physics was created to describe electron orbital behaviour based on wave theory. Light was then linked to electron energy levels and orbital transitions through a new concept called quantum physics. A quantised particle called a photon was proposed as the quantised energy light particle related to these energy transitions.

LIGHT AS WE KNOW IT - ELECTRON PHOTONS (EP)

Physics theory proposes that light is made up of quantised energy "particles" called photons. There are many types or "sizes" of photon, or perhaps I should say wavelengths, which make up a continuous (quantised?) spectrum of light. The energy and wavelength of light particles depends on how they are created based on quantum theory. It states that the energy of a photon is related to the change in energy of an electron (electromagnetic field) that creates it. The amount of energy transferred (lost or gained) is the amount of energy contained in the photon.

Photons created by electron energy level transitions have been studied and analysed extensively by physicists. In simple terms an electron is energised and then de-energised to make light. An amount of energy is captured by an electron (somehow) and when spontaneously released (somehow) it creates a quantised light particle or photon. When matter is heated and its atoms and hence electrons gain more energy, light is also created. In this process, energy added to an atom causes quantised electron energy increases or jumps in electron "orbits" (shells?). This quantised energy may then be released in a reverse transition creating light. Electron energy transitions can also be stimulated in electrical circuits, which is the basis of modern radio technology. Figuratively electrons are "rubbed together" or heated to make light not fire.

The photons we "know" or have learnt to live and work with in almost all areas of science and technology are created from electrons. To help with understanding I have created new terminology for such light and called photons created from electrons "electron photons" or "EP" for short. There is a wide and supposedly continuous spectrum of EP light created by electron activity ranging from very low frequency, long wavelength, low energy light created by almost "mechanical" electron manipulation, through optical wavelength light and up to "very high" frequency, "very short" wavelength light. I put very short and very high in inverted commas because these are somewhat subjective measures. Very short wavelengths include X-rays and Gamma rays which are supposedly associated with very high energy electron activity.

The light creation and capture process and the associated energy or wavelength is also supposedly always fixed for any given type of atomic electron energy level transition. This is the basis of atomic line spectra which are discussed later. But supposedly this process may also depend on other factors such as the speed of the source or at least the relative speed difference between the source and detector. The energy of a photon also supposedly remains constant for the life of the photon. But it has also been proposed that energy levels may also change over time and distance once a photon is created (gradual decay, cosmological expansion, etc.).

Many sources of light generate many different wavelengths or colours. In the visible spectrum, white light is made up of many wavelengths or colours or many "sizes" of photons. The relative mix or spread of wavelength determines "whiteness". For example the white light we know best from the sun has an almost flat spectrum of light over a wide bandwidth. This ranges from red light up to blue light and even beyond at both ends of the spectrum. Slight differences in the spread that may or may not be detectable by the eye, can make other types of light. Sometimes this is called "pink light".

Light from specific types of atomic activity, such as laser light, is monochromatic and produces very narrow line spectra. All light (all photons) from these sources has exactly the same (or very similar?) energy per photon and hence wavelength. However, there may still be a small spread in the bandwidth of such spectral lines (e.g. a 3 dB B/W) and this is a challenging issue.

The table 4.1 below shows electron and some electron photon (EP) energy levels and wavelengths derived using simple laws of physics (mostly in scientific notation.)

PARAMETER	UNIT	PARTICLE	PHOTON	PHOTON	PHOTON	PHOTON	PHOTON	PHOTON
		ELECTRON	MHZ	GHZ	THZ	visible light	xrays	gamma rays
FREQUENCY	HZ	1.24E+20	1000000	1E+09	1.00E+12	1.00E+15	1.00E+18	1.00E+21
WAVELENGTH	metres	2.43E-12	300	0.3	3.00E-04	3.00E-07	3.00E-10	3.00E-13
ENERGY	Joules	8.19E-14	6.63E-28	6.63E-25	6.63E-22	6.63E-19	6.63E-16	6.63E-13
MASS	Kgm	9.10E-31	7.37E-45	7.37E-42	7.37E-39	7.37E-36	7.37E-33	7.37E-30

TABLE 4.1 - ELECTRON PHOTON (EP) PROPERTIES

But is there an upper limit to the wavelength or photon energy of electron light? How much energy can an electron be made to capture and then release in any one transition or "jump"? Can it release more than it can contain? According to current theory the "total" energy of an electron is about 1E-13 (10 to the power minus 13) Joules, where one Joule is roughly equivalent to 6 * E+9 GeV. The energy of a Gamma Ray photon is approximately 7 E-13 Joules (about +4 MeV) which is more than the total rest energy of an electron. Perhaps if an electron is accelerated to near light speed it can gain more energy in terms of an increase in mass. This high energy when released in a collision or annihilation may be sufficient to create such high energy photons.

So we now know that light comes from electron activity. But what other forms of light creation or capture could there be. It is well known from quantum theory that all particles and bodies have wave like behaviour. Even the sun supposedly has a wavelength associated with it. So is it possible that light could come from other atomic particles or can it only come from charged ones? Is light simply the result of changes in the energy level of any dynamic charged particle? If so, then what other particles, especially charged particles, could be "rubbed together" to make light? Of course there is only one other atomic option, quarks excepted.

DUALITY – WAVE, PARTICLE OR BOTH

The next property or perhaps mystery of light is the so called duality property. Light behaves in two ways or has two properties at the same time, that of a wave and that of a particle. It seems schizophrenic or perhaps to put it more politely suffers a bipolar disorder and is not sure if it is a wave or particle. Or perhaps it does know and it is just our understanding that is at fault. There is an interesting parallel here in mathematics between the use of the time domain (particle) and the frequency domain (wave) in Fourier analysis but I will save that idea for further study.

THE WAVE

The idea of a "light wave" (or is it light waves) was created to match the interference results that were observed. The similarity between the interference results of light and the interference behaviour of wave forms such as sound waves in air, waves in water or other fluid wave movements, was striking. So it was assumed by even the best physicists that light was carried by some form of wave. This is the basis for the wave theory of light.

The wave behaviour of light is based on the behaviour of longitudinal waves and follows a similar mathematical theory. But that is where the similarity ends. Unlike most other well-known waves, light "waves" have no longitudinal dimensions. From a simple analysis of the properties of light, a light wave can only be a lateral wave. But if it is only a lateral wave for each photon, is it really a wave in the common understanding of waves or is it a wave in terms of behaviour only? So then how does wave behaviour relate to light being made up of many unconnected or unrelated photons or are they related in some unknown manner?

When light is quantised and a photon is "created" these particles seem inconsistent with wave theory so the idea of a "wavelet" was proposed. Initially a wavelet was thought to be made up of a bundle of photons. The question then became how are individual photons "connected" together to form a wavelet. Any answer to this question created problems with the photon creation and capture processes. It required a mechanism to connect what were believed to be asynchronous and random photons. The next idea was based on each single photon somehow becoming a longitudinal wavelet. But this also created difficulties as explained below.

THE PARTICLE - PHOTON

The particle properties of light are based on experiment and theory. The work of two great physicists, Plank and Einstein, eventually solved the issue of whether light was a particle or wave, or did they? It was proposed that discrete packets of energy, not a continuous flow of energy was how light must behave. This was seen as the only realistic solution for the ultraviolet catastrophe and for the photoelectric effect. The light quantisation theory earned Plank a Nobel Prize in Physics while the photoelectric theory earned Einstein one. Light quantisation has been "verified" by many experiments and fits in well with or should I say was the start of quantum theory. But perhaps these theories added more questions than answers about what light is.

At last a light "particle" had scientific support. This particle was eventually given the name photon based on "photo" for light and "on" as a suffix for a particle, just like the electron, proton and neutron. Each photon was proposed as being made up of a discrete quantised "bundle" of E/M energy. Current theory directly relates the energy of a photon to the change in the energy level of the electron which created it. A photon is somehow "created" when energy is lost by a field of an electron and the reverse process occurs when it is "captured" and the same amount of energy is regained by the field of an electron. I have used inverted commas because I believe the full process of photon creation and capture is not well understood by even the best physicists. This process may be described in quantum mathematical terms but may not yet be explainable in simple physical or should I say in "mechanical" or perhaps "electro-mechanical" terms.

Each photon is supposedly made up of a mutually inducing orthogonal pair of "bouncing" electromagnetic vectors, an electric field and a magnetic field. Each bouncing lateral field creates a longitudinal wave effect when viewed from any position orthogonal to the direction of motion. These two fields form a composite vector field like a "rotating flat disc" that gives the direction and magnitude of the light. The composite field travels in a virtual helical spiral in the direction of travel and creates a composite wave effect when viewed from any longitudinal position. This model satisfies the quantum requirement for light and the Special Relativity requirement for zero longitudinal dimensions (addressed below) but what are its lateral dimensions and how does it move forward and also contain energy and mass?

A photon can't have infinite lateral dimensions as this has unacceptable energy density implications and requires infinite peripheral speed. Both the electric and magnetic fields can't gradually thin out and have no boundary. Therefore each electric and magnetic field vector and the photon itself must end completely at some type of boundary or edge. These two fields and the photon disc must have no energy beyond this point. The energy density across the disc may vary but the disc size must be fixed. The finite photon disc size or diameter must be related to wavelength (dia. $\sim \lambda/2\pi$) which is inversely related to frequency and hence energy or "mass".

A new parameter "α" is proposed for photon disc size or "disc area" (capture area). It would be related to the square of the wavelength ($\alpha \sim l^2$) and hence inversely to the square of the frequency. The energy density of each photon disc would therefore be inversely related to the wavelength which is important for the capture process. If a photon (disc area) is of a particular size (i.e. wavelength), then it can be "seen" by an object that has that "size". Therefore if the size of the photon and the capture device are the same or perhaps just wavelength related then the photon may be captured and an energy transfer occur. If a photon is not the same size as the capturing device, especially larger than this "capture area" it would be undetectable.

It is interesting to note that a longer wavelength photon which has lower energy would have a bigger "disc size" and hence much lower energy density than a shorter wavelength photon with more energy and a smaller disc. But is the energy evenly spread out over the disc? The total energy must be fixed for a given wavelength photon but the energy density across the disc may be variable. It would be a function of the composite field strengths of the two fields and may possibly be a function of radial distance. This combination may produce an inverse cubic law for energy density as a function of radial position and as a function of wavelength. Of course these are just proposals that require further work to explore, derive and prove in experiments.

Polarisation or orientation and other factors need to be taken into account in the capture process. This doesn't change the size requirement just the alignment. Polarisation can readily be explained using the "rotating" flat disc model. The fields for a photon are always orthogonal and are effectively fixed in orientation once created unless influenced by an external field or media. If the photons from a source are aligned so that the angle of either field viewed in the direction of travel is always or predominantly in the same direction then the light is said to be polarised. If the photon fields are randomly aligned then the light is not polarised. The degree of polarisation depends on the degree of alignment of the fields. Polarisation may be changed in particular situations such as traversing some types of media or electromagnetic fields.

So each quantum of light or photon is like a spinning virtual flat disc with no thickness in the direction of travel. It spins orthogonal to the direction of travel and is "observed" to move at the speed of light in the direction of travel. But perhaps a photon has "no knowledge" of movement if that is an appropriate term. It can't "see itself" moving forward or travelling in any real sense as it has no dimensions in the direction of travel. Perhaps this is really why every observer sees photons as having the same speed. To an extent, perhaps the observation itself may be creating the third dimension of the photon and hence defining its speed.

SO WHICH IS CORRECT?

This apparent schizophrenic nature of light has puzzled scientist for a long time. The debate over which interpretation is correct or which better explains the various behaviours of light continues to rage. I remember asking questions about this duality "problem" when I was a young physics undergraduate student. It was explained to me that a photon was something like a small wave packet or wavelet. But if light really is made up from little wave packets or wavelets how do these relate to particles. Is a wavelet made up from one photon or many and if so then how many? I am not sure how the wave/packet duality nature of light is explained today or how a photon based wavelet is modelled but there are some key aspects that need to be considered.

For light to be a longitudinal wave it must have longitudinal dimensions. Light is proposed as being made up of wave-packets with longitudinal dimensions and properties such as phase velocity, group velocity and packet speed. But how can light be both a longitudinal wave and a "flat particle"? While longitudinal wavelets fit well with the conventional wave theory of light, how do individual photons fit in? And the wavelet explanation also has longitudinal implications which are problematic. So let's take a closer at a photon to get a better understanding of it.

A photon can't have any longitudinal dimensions at all. A simple analysis based on Special Relativity shows why. It comes from inserting the speed of light as the speed of the reference frame in question in the Lorenz length transform. The Lorenz transforms imply that anything travelling at the speed of light must either have infinite longitudinal dimensions hence infinite length or zero longitudinal dimensions or no length. Now for a finite energy particle such as a photon to have infinite dimensions implies zero average energy density. On the other hand finite energy density and infinite length implies infinite energy and these are meaningless outcomes.

Therefore light "particles" or photons which are travelling at light speed must have no dimensions in the longitudinal direction of travel. A photon must therefore only be a two dimensional "particle" or lateral object, a type of "flat disc". It can't be a longitudinal particle or "wave". I am not the first to note this interpretation of zero longitudinal dimensions for light particles. It had been realised by physicist since before the time of Special Relativity. But it still seems to be misunderstood and not well applied in discussions of light behaviour. Wavelets with longitudinal dimensions are still commonly used to explain light behaviour.

Perhaps a more realistic proposal for light is that it is neither a longitudinal wave packet nor a conventional three dimensional particle. It is simply a discrete, modular (quantised) fixed size two dimensional photon discs of coupled fields. Each photon disc is made up of two orthogonal one dimensional finite fields, an electric (E) field and a magnetic (B) field. Each field oscillates back and forth laterally, mutually inducing the other. When one field is at a maximum strength it has maximum acceleration but zero rate of change so the other field is at a minimum or zero point. Together they create a composite vector, similar in a way to the spinning magnetic field of an AC (synchronous) electric motor but perhaps I am showing my engineering background here. The vector cross product of the two fields (E and B), called the Poynting Vector, is orthogonal and gives the magnitude and direction of the light.

This model satisfies the quantum requirement for light and the special relativity requirement for zero longitudinal dimensions or length but questions remain. In particular, what are the lateral dimensions or size of the so called light disc and how does it contain energy and mass if it has no volume? Each photon contains energy and momentum and can be said to have mass but not matter or mass in a conventional sense as it only has two dimensions. The dimensions and other factors such as radial energy density must be related to the wavelength but more analysis is required in order to model and determine these parameters and estimate their values.

This proposal of a photon as a flat disc is an interesting area for further study, especially in relation to the analysis of light behaviour in general. Polarisation can be easily explained using this "rotating" but "fixed orientation" flat disc model. Is this finite size flat disc discrete photon model an appropriate way to analyse sother type of light behaviour? There may be areas of light behaviour such as interference that can be more easily and accurately explained using this model. This is the approach I have taken in my analysis of heavy light and some of the light behaviour ideas expressed in the following chapters.

But are there more sides to light than we know about? Does light have an even more complex psychology to continue the analogy? Can light, especially individual photons relate to each other in other ways? Individual photons in general don't seem to be related from the time they are created, apart from perhaps having similar properties such as wavelength and polarisation. But can separate photons somehow combine to behave in a collective and even longitudinal manner? Do photons ever combine together to make a composite wave or does each individual photon just wave as it goes by? Can they be related in a time or space by having fixed separation due to some form of synchronisation in the creation process or due to some intermediate media or field effect which causes some type of synchronisation?

Can two photons be related or connected so that one can cause a behaviour change in the other. There is a quantum property along these lines called entanglement. This is supposedly responsible for some strange light (and other quantum?) behaviours. It is said to apply to particles which are somehow connected or related during the creation process or through some mutual media transmission process. In particular it is said to apply when a photon supposedly splits into two parts (two photons?) during diffraction, creating two photons that are "entangled" or related to each other in some way. I have doubts about photon splitting and entanglement. It seems like epicycle material to me. But is there more to inter-connected photons?

SPEED OF LIGHT

One of the properties of light that challenged early scientists was how fast light travels? Perhaps the earliest view of the speed of light was that it was infinite. But infinite speed was inconsistent with real physics. So it was realised it must be finite but still considered to be very fast and possibly unmeasurable. And was the speed of light the same for all types of light (colours or wavelengths)? Does it change under certain conditions or through different media and how could this be measured? These questions challenged the best scientists at the time.

Perhaps the first reasonably successful measurement of the speed of light used very large cosmic distances, unlike Galileo's short earth bound distances. It was observed that one of Jupiter's moon phases was longer when it was further from the earth compared to when it was closer. This fascinating observation was amazingly and correctly proposed to be due to the difference in the transit time of light for the two positions involved. This led to an estimate for the speed of light of about ***2.2 *10⁸ m/sec***, which was surprisingly good given early planetary distances.

At last the speed of light was captured so to speak but more accuracy was required. This started a race for better methods and the focus shifted to earth bound experiments. One of the earliest methods used a mechanical device called an interferometer. This had a large spinning drum with slots regularly located around its periphery. Light was shone through the slots and the rotation speed was varied until a synchronous interference pattern was detected. The speed and dimensions of the wheel were used to derive the speed of light and a figure of around ***2.8*10⁸ m/sec*** was determined. This was verified many times and is very close to the mark.

Experiments based again on rotating devices but using half mirrors to create longer paths, finally increased the accuracy of measurement to about one percent. For the relatively crude mechanical devices and measurement systems used, this was a reasonable accurate and readily repeatable result. But while more accuracy could possibly be obtained with better equipment, there were fundamental problems with this method. There were limits on the accuracy of distance and time measurements that were used to produce a result. These set limits on the accuracy of the speed of light that could be derived using simple mechanical methods.

The next big step in determining the speed of light was theoretical and Maxwell's equations did the trick. These showed the speed of E/M waves (light) was a function of other basic parameters of the media. These parameters, the Electric Permittivity and the Magnetic Permeability of the media in which the light travels, or more correctly in which the two coupled fields oscillate, could be measured or obtained from other experiments.

The Electric Permittivity, represented by the Greek letter ε describes how an electric field affects a dielectric medium. It indicates the ability of a material to allow (permit) an electric field to penetrate it and is determined by the amount a media is polarized or electrically aligned in response to a field to reduce its strength. It is measured in farads per metre (F/m). The permittivity of free space, ε_0 (also called vacuum permittivity or electric constant) is given by;

$$\varepsilon_0 = 8.86 \times 10^{-12} \text{ F/m.}$$

The Magnetic Permeability, represented by the Greek letter μ, is the degree of magnetization of a material in response to an applied magnetic field. This term is measured in Henries per metre, or Newtons per ampere squared. The constant μ_0 is known as the magnetic constant or the permeability of free space (vacuum), and has the exact or defined value of;

$$\mu_0 = 4\pi\times10-7 \text{ N/A}^2.$$

From these two parameters and using Maxwell's equations, the speed of E/M waves or the speed of light c, is given by;

$$v = c = 1 / \sqrt{}(\varepsilon * \mu)$$

In a vacuum with the values of ε_0 and μ_0 this becomes,

$$c = 299866 \text{ Km/sec}$$

This theoretical derivation of the speed of light is very important in physics. It shows that the speed of light is a function of two fundamental electromagnetic properties of the media or space through which it travels. This equation also implies that light speed is not absolute or constant but changes as the parameters change. If either parameter is increased then the speed of light decreases and vice versa. Interestingly there seems to be no obvious reason for an upper speed limit from this equation and properties. If either could ever become zero then the speed of light would be infinite and if either becomes very large then the speed of light becomes very slow. Perhaps apart from their finite fixed values in free space, there are limits to these properties for any particular state of matter.

But for one special state of matter called a Bose/Einstein condensate, light supposedly can be slowed down (to zero??) due to "quantum effects".

Apart from the obvious media dependency on ε and μ. what other parameters, fields or physical effects could change the speed of light? Einstein's Special relativity "proved" that light speed was constant and the same for all inertial observers. But in General Relativity he proposed that it can change in "gravity fields" although the change was again supposedly the same for all (inertial and non-inertial?) observers. This behaviour was supposedly measured using star shift observations near the sun during a solar eclipse. Such a possible change in light speed due to gravity was first derived using a conventional concept of photon mass and the law of gravity.

The effect of gravity on the speed of light is also behind the Big Bang theory, Big Black Holes and many more Big Ideas. But does this really happen or is it an elaborate theory used to explain some observations that have yet to find another cause? There may be other explanations for such light bending phenomenon which perhaps have not been properly allowed for. So is light bending by gravity another "epicycle" helping to save G.R. and the Big Bang Theory?

Today the speed of light is derived using lasers and atomic clocks. Current methods using the latest high technology optical devices and solar measurements have reduced the error down to 1 in100 million. The SI standard speed of light is now 299792458 metres/sec. This speed of light is now used for measuring distance, not the other way around as before. The standard metre rod which was used for measuring distance has been replaced by the speed of light and a fixed time period but the rod is still more manageable for every day applications. I have used an approximate light speed of 3 *10E8 m/sec for general free space light speed calculations.

MASS – ENERGY, MOMENTUM AND PRESSURE

This is not intended as a pun but one of the heaviest mysteries about light is does it have mass or weight, and if so, how much does it weigh? Weight is a gravity based concept but it is a useful introduction. Mass, or the amount of matter, is what is really important for a particle. But how can light contain mass or matter? The other question about light that has challenged physicists for a long time is how does it react in "collisions"? Are collisions always "elastic" with no energy transfer or is light captured and re-emitted in all or some atomic collisions? Does an "elastic" collision always transfer momentum and does this process always cause pressure?

From the work of Maxwell, Plank and Einstein, light in its quantised form of photons is shown to have energy related to its wavelength or frequency. This energy then implies momentum and therefore an equivalent mass. Each photon has energy and equivalent mass again from Special Relativity. The equations that provide these figures are given below.

$$E = h * \nu$$

Where E is energy, **h** is Planks constant and ν is frequency. And also

$$E = m * c^2$$

Where m is "rest" mass of a photon and **c** is the speed of light. So,

$$m = (h * \nu) / c^2$$

$$\text{or} \quad m = \nu * (h / c^{2)}$$

Where h/c^2 is a constant, approximate value $7.7 * 10^{-50}$ (which is a very small number).

Therefore the momentum of a photon, $M_{p,}$ is given by;

$$M_p = m * c$$

$$= ((\nu * h) / c^{2)} * c$$

$$= (\nu * h) / c$$

$$\text{or} = \nu * (h / c)$$

Now h/c is another constant, approximate value of $2.2 * 10^{-42}$. This is also a very small number but when applied to very high frequencies and to a very large number of photons it produces measureable outcomes of momentum. The momentum of light or of each photon is readily calculated from a simple analysis of photon energy, equivalent mass and the speed of light. Pressure can be derived from the transfer of momentum in collisions.

Maxwell was one of the first to come to the conclusion that light causes pressure. It came out of his work on electromagnetism. This was later tested by Russian physicists Lebedev in experiments early last century and confirmed shortly after by English physicists Nichols and Hull. Of course the momentum of a photon is related to energy and hence frequency or inversely to wavelength. The shorter the wavelength of light the greater the frequency and hence the greater the energy and momentum if has. So very high energy, very short wavelength light creates greater pressure per photon in elastic collisions.

For a complete elastic collision of a photon at right angles to a "hard surface", this transfer of momentum is turned into a pressure, P, where;

$$P = \rho * r * A$$

$$= \nu * (h / c) * r * A$$

$$= \nu * r * A * 2 * 10^{-42}$$

Where ρ is the momentum of a photon, **r** is the photon rate per unit of time and per unit of area, **A** is the surface area of the body and ν is the frequency of the photon.

A body exposed to light will experience a force or pressure on it from that light. If the light is absorbed the pressure is the power flux density divided by the speed of light. If the light is reflected then the pressure is doubled. Of course if the light is absorbed and turned into some other form of energy which is then re-radiated, this may create a reactive pressure or a type of negative pressure effectively matching the initial impact pressure. If light on a body (spherical?) is iso-cosmic, there is no net pressure difference and hence no net force on that body. However, if the light intensity is directional in strength then there would be a net directional force.

The sun creates light pressure on earth and every other planet. This is the basis of the proposal to use solar sails to drive a spacecraft. Light pressure from the sun from electron based light is however very low. The amount of pressure depends on its size and distance from the sun. For example the flux density of sunlight on earth, assuming the energy is fully absorbed, produces a pressure of about 5 E-6 Pascals. If it is totally reflected it is twice this or about 1 E-5 Pascals. The surface area of the earth is about 10E+14 sq. metres so this produces a net force on the earth by the suns light pressure of order 10E+9 Newtons. This is insignificant compared with the "attractive" force of gravity from the sun of order 10E+22 Newtons. Another source of electron light pressure on earth is from cosmic background radiation (CMBR). This depends on the wavelength and intensity/density of CMBR but is also insignificant.

Some forms of electromagnetic radiation and hence light pressure are obvious. But are there other forms of light or cosmic background radiation that are just as important but not easily detectable? Perhaps other types of light are involved in a lot more behind the scenes activities than we realise. The question of the relationship between light and gravity has fascinated me for a long time. If there are other sources of background radiation like cosmic gravity background radiation (CGBR) as per my gravity theory then the pressure may be significant. This is the basis of my theory of light gravity and drove me to develop the theory of a new type of light.

In calculating light pressure there also needs to be factors which allow for the degree of penetration, the rate of reflection or absorption and the angle of incidence and the angle of reflection of photons. These factors could reduce the pressure by around a factor of 2 each and together less than an order of magnitude. They would reduce the overall pressure of sunlight on earth and pressure from other sources such as CMBR but it is still insignificant in any real terms. These factors should be taken into account in deriving light pressures from light that produces gravity. But it probably produces less than one order of magnitude reduction.

However, the rate of gravity light or CGBR and the wavelength or energy per photon are still unknowns. It may be possible to approximate the density or rate of CGBR from CMBR data assuming it is a similar type of universal background radiation. But deriving the energy per photon is a much more complicated problem. It requires an estimate of the frequency or wavelength and other parameters to enable an estimate of light intensity that causes gravity pressure to be determined. These factors could also be estimated in reverse using current values of gravity force converted into a required pressure difference. Unfortunately this requires a method for deriving the pressure differences caused by shadowing and that formula is still unknown. This is addressed in the chapter on my new light but is also for further study.

SPECTRUM and RED SHIFT

White light is made up of many wavelengths or colours so the photons can have many "sizes". The mix or spread and relative amounts of light determine the "whiteness". For example the white light we know best from the sun has a broad and flat spectral composition due to its high temperature. On the other hand photons from monochromatic light such as a laser have (exactly?) the same amount of energy and the same wavelength thus producing "very narrow" line spectra. The question about the width of line spectra is discussed later.

It was observed that different elements gave off different colours of light when heated. It was also found using a prism to create a spectrum, that light emitted from a hot body was not always continuous. There were different patterns or lines of light at discrete frequencies creating line spectra. There were in fact different line spectra for different materials. The property was also noted to occur in reverse when white light was shone through a gas and absorption spectral lines were observed. These were noted as being the same wavelength of the emission lines generated by heating the same type of gas but had the exact opposite effect of producing black lines in an otherwise continuous white light spectrum.

Observing spectral activity provided a means of identifying elements but the real cause or reason for the spectral line effect was still not clear. An explanation was eventually proposed and it was tied up with the planetary atomic model. The electron "orbits" were found to be related to electron wavelengths. This added to the understanding of electron energy levels and the development of atomic and quantum theory. This connection between atomic activity and light line spectra was a major discovery but many questions remain. Do all photons in a specific line spectra have the exactly the same energy and wavelength or is it distributed like a Bell shaped curve? Are all spectral positions occupiable or are they also quantised as per quantum theory with only certain positions or energy levels allowable? (Like integers on a number line)

Spectral theory was important to the development of many new areas of physics, in particular, the study of distance in cosmology. It was found that light from some very distant sources (very distant at the time?) had spectral patterns which aligned with known elements only if they were shifted slightly in one direction. The spectra of Hydrogen and Helium were observed from many cosmic sources but the patterns were at slightly different wavelengths. A relatively small shift easily realigned the line spectrum with ones from local or reference spectral sources. The amount of shift was found to be the same for all spectral lines observed from a given source. The shift was not a function of wavelength but was found to be linear across the entire spectrum.

The parameter used to measure cosmic light shift is called Z. It is based on a comparison of the wavelength of the shifted light compared to a reference source. It is a dimensionless quantity defined by the following equation;

$$1 + Z = \lambda_{shifted} / \lambda_{reference}$$

where;

$\lambda_{shifted}$ is the wavelength of a specific line from observed light.

$\lambda_{reference}$ is the wavelength of the same line from a local (reference) source.

If the spectrum of the received light has to be shifted toward the red end of the spectrum to re-align it with a reference source, it is called red shift (has lower energy) and if toward the blue end it is called blue shift (has higher energy).

There is now a whole new field of cosmic observation and research related to light shift observations. This then triggered another of field cosmic research associated with the model of the universe. It started the expanding model of the universe (which was also supported by G.R.) and produced the Big Bang theory of the origin of the universe and "Black Holes" theory.

SUMMARY

This chapter provides a basic platform of measureable parameters (set of metrics) or at least properties which new light theories can be examined against. In the chapters which follow a range of new ideas on light theory are explored. In some cases the above photon based properties are used as a means to model, understand and try to explain them. This work is very initial and exploratory at present but should provide sufficient background to support my ideas.

All new ideas on light should eventually undergo rigorous theoretical analysis to verify them. New theories need to be analysed in mathematical detail using appropriate quantum theory and supported quantum parameters to prove or disprove them. But I am not yet in a position to do this work and leave that to more qualified theoretical physicist to consider. Hopefully this book will trigger work in this direction and I look forward to monitoring progress and even participating in it. Any contributions (constructive?) in this regard would be appreciated. In the meantime let's take a closer look at some of my new ideas about light.

CHAPTER 4 - A THEORY OF NEW LIGHT

This chapter proposes, defines and examines a new type of light. It is the theme of this book and is the main new idea about the physics of light that I am proposing. This new light is also behind my new gravity theory as set out in my first book. While other ideas on various aspects of light behaviour addressed in this book may change the way we see light, this new light will have the greatest impact. It will open up new fields of physics for exploration and change many areas from atomic and quantum physics up to cosmic physics and the way we view the universe. Perhaps the physics of light, as well as the theory of gravity are about to undergo another amazing leap forward. I am certain that many new fields of science and technology will emerge from this discovery and hopefully they will produce many new benefits for mankind.

This new light is examined in terms of the properties of light set out in the previous chapter. It is a high level review rather than a formal mathematical analysis and like most areas of physics is a work in progress. But it is scientifically based and all analysis and proposals are logical, objective and practical. Perhaps it is not formulated in sufficient mathematical rigour for the experts but is explained in enough detail to launch it as a sound scientific proposal and initiate further consideration. I hope that this initial work will set the scene for more experimental research and in depth analysis by relevant experts. This should then lead on to more formal proof and a detailed theory about my new light. Hopefully it will be proven to be correct.

So what is so special about this new light? What are its properties and what does it do? In particular, does existing light theory apply or is it different from the light we know?

Specific questions that immediately come to mind are:

- Where does this new light come from?
- What are the wavelengths and energies?
- What is the speed of new light and is it "c"?
- Do Maxwell, Plank, Einstein and other light theories of apply?
- What is the energy, mass and momentum - light pressure?
- How does it behave - refraction, interference, red shift, etc.?
- How can it be created/captured, observed and measured?

THE SOURCE OF LIGHT AND THE ATOM

Perhaps one of the most important questions is what is the source of light? To answer this question requires a closer look at the atom which plays such an important role on all aspects of physics and especially light. Atoms are part of everything we touch but they play a critical role in creating the light that lets us see all things. They also create "light" at other non-visible wavelengths. But how does it all happen within the atom from atomic component activity.

Current atomic theory proposes that light is simply the result of energetic stimulation of a negatively charged particle, the electron. All wavelengths, all spectral lines and all light energy that we know about is somehow the result of electron activity. But are there other ways of making light? What other energy sources or particles can be "rubbed together", so to speak, to make light? Can any atomic charged particle make light? These challenging questions are what led to my idea of a new type of light. But what is this new light? Where does it come from and what are its properties. More specifically what is its real source and how is it generated by atomic activity? And what does it mean for the components and structure of the atom?

But before those questions can be addressed we need to revisit the atomic model, in particular the nucleus and the proton and see how they behave from a more mechanical point of view. In quantum theory all particles supposedly have wave like behaviour. Even the sun has a (very short) wavelength associated with it. But does this wavelength concept apply to all atomic particles in a similar way or is this purely a theoretical creation used to develop the electron model? For example, can the proton behave in a similar way to the electron? Can it also have wave like behaviour? Yes I can already hear the quantum physicists saying they are different types of particle, one is a lepton and one a hadron and different quantum rules apply? But I won't get into that debate just yet and will continue with a more mechanical or perhaps conventional discussion about atomic particles and how they may create light photons.

It is well known that there are two charged particles in an atom, the electron and the proton. Both have exactly the same charge quantity (which in itself is amazing) but they have opposite polarities (whatever polarity really means). The electron has been given a negative polarity and the proton a positive polarity but this is purely by convention. The real meaning of the property called "charge" as well as polarity, are not well understood I believe. It is simply by definition that there is a basic unit of some type of "electrical property" which can occur in equal (quantised) amounts and with opposite polarities on these two subatomic particles. But what other differences are there between these two particles? While the electron is supposedly indivisible, current quark theory proposes that the proton is divisible. Quark theory also proposes fractional charge amounts. But these aspects are still primarily theoretical and subject to conjecture so I will proceed without specific reference to them for now.

The only other property of atomic particles that seems important in any discussion of activity and light is mass. Current theory proposes that the proton and electron have mass and the proton is several orders of magnitude more massive than the electron. This means it has more energy according to mass/energy equivalence. This increased energy has implications in terms of the dynamics of the proton. While it may be heavier than an electron, it can still move about but any movement must have significantly higher energy implications. A unit used for the energy of an electron (and other particles?) is the electron volt (eV). Perhaps a new unit called the proton volt (pV?) could also be established. If it is based simply on electrical potential the two units should be equal. But if it somehow includes the mass or energy of the particle in some way, the proton volt would be a much greater energy unit.

Current atomic theory proposes frequency is directly related to energy (Plank) and that energy is directly related to mass (Einstein). Therefore if mass is increased then energy is increased and frequency is increased or wavelength is decreased. Quantum theory also shows that mass and energy are quantised and hence wavelength is also quantised. So what does this mean for the proton? The proton and the electron have similar but opposite electrical properties but a significant difference in mass or energy and hence in "size". Therefore a proton which has much higher energy must have a much higher frequency or shorter wavelength than an electron. But how do these properties impact on behavioural differences or the dynamics of each and on light creation and capture and photon characteristics or parameters?

It would seem that the similarities between the charges of an electron and a proton (equal in magnitude but opposite in polarity) should make them behave the same electromagnetically. Therefore Maxwell's equations which were originally developed for an electron and associated electrical fields and their related magnetic fields (general E/M fields) should also apply to proton activity. The theory initiated by de Broglie for determining electron orbits based on wavelength of an electron should also apply. I agree this connection is not based on sound quantum theory and that will need to be developed in due course. But I believe the general behaviour of electrons and protons should be similar apart from appropriate scaling for mass/energy differences.

Therefore as the electron and the proton are charged atomic particles that "exist" in the atom, both should be capable of being involved in similar quantum atomic actions such as dynamic movement and energy level jumps and hence light (photon) creation. This is where a real leap in science is proposed. Protons should be capable of creating light or photons, albeit at much higher energy levels and hence with much shorter wavelengths. End of story! I just hope this statement (bold scientific theory?) doesn't upset too many physicists. But before proceeding too far with this bold new idea I will re-examine the atom or at least the model of it and try and evaluate what this new dynamic photon theory means to the atomic model and light.

NEW ATOMIC MODEL AND HEAVY LIGHT

What are the implications of this dynamic proton proposal on the atomic model? One feature of an atomic model that doesn't seem to have been fully recognised, at least as far as my research into the literature goes, is that an electron can't simply orbit a "solid", stable, static nucleus. The picture must surely be much more complicated than that. The nucleus or at least the protons cannot be fixed and static and "held in place" by some (unknown) nuclear force. All charged particles, including the proton must be active or dynamic in order to keep opposite charges apart and not let them merge and keep like charges together and not let them fly off in all directions.

For example, let me use a cosmic situation (without charge) to try and explain what I mean. Many people believe the moon orbits the earth which is correct in a sense but in another way it is not. In reality both the moon and the earth orbit each other. In a way it could be said that the earth is in a geostationary orbit of the moon because we always see the same moon face. But to put it more correctly both orbit about a common centre of gravity. This common centre of gravity is determined by the masses and separation distance of the bodies and is much closer to the earth than the physical mid-point. The earth also orbits this centre of gravity (C.G.) and the combination of the two bodies acting through this CG orbit the sun. It could be said that the earth "wobbles" as it orbits the sun. The wobble is quite slow (28 day periodicity) but it may be noticeable in cosmic observations. It would have implications on cosmic distance and other measurements but I am sure it is already taken into account in all such cases.

This concept can be extended by looking at a simple Hydrogen atom consisting of one electron and one proton but no neutrons (only rare isotopes of Hydrogen have neutrons). Charged particle attraction and perhaps gravity as well as other quantum aspects are involved but this shouldn't change the general situation. Following the above concepts, the electron and proton in a Hydrogen atom are in some form of mutual "orbit". The electron can't just simply be orbiting a fixed proton. They must both be orbiting each other or more correctly orbiting about a common (electrical/mass?) "C.G." point. In the case of hydrogen this point would be very near the proton and it could be called the "centre of the atom". The proton must therefore not be "stationary" but also must effectively "orbit" around this common central point.

The basic concept of "orbiting" protons for Hydrogen atoms could then be extended to other atoms even those with neutrons. This type of balanced movement must apply to all electron–proton pairs in all atoms. It would seem that each electron/proton pair in any neutral (non-ionised) atom must be effectively orbiting each other. Of course each electron may not be connected with a specific proton and cloud quantum aspects would most likely apply. But this shouldn't change the concept of orbiting protons. A dynamic proton must have wide implications. For larger atoms in general there may be many local "C.G.s" or an overall one. There may be an overall central point somewhere in the "bunch" of neutrons that could be the effective "centre" of the atom. Where this point or these points are is perhaps not significant in this general discussion but may have significance in a more complete picture. In any case the proton is surely an active particle. This proposal is scientifically sound but needs more proof.

This similarity in general atomic behaviour between electrons and protons must surely be the correct explanation of atomic structure. In particular "orbiting" protons must have energy levels as per "orbiting" electrons as well as other quantum behaviours. The exclusion principle, uncertainty theories, energy levels and transitions that have been developed for electrons must also apply to protons. This new dynamic proton must have major implications for the old (current) atomic model, especially the "solid" nucleus. This behaviour needs to be interpreted in terms of the de Broglie model of orbiting electrons and the current quantum theory probabilistic model of the atom and its components. A whole new field of atomic physics will need to be developed for the proton including energy levels, states and transitions and other quantum parameters in a similar way to that developed for the electron.

So all atomic protons must also be "orbiting" or moving (shades of Galileo, "and yet it moves"). How profound! The key point is that the nucleus is not "solid" and electrons don't just simply orbit a fixed static nucleus. Protons must be in a similar type of orbiting dynamic, appropriately scaled for energy differences and hence much closer together or smaller than for the electron.

ORBITING PROTONS – PARKES (2005)

It is well known that the concept of an orbiting electron in a classical planetary sense is only that, a concept. But that concept was used successfully to explain the process of electron stability, energy levels and of course light creation and capture. Current quantum theory proposes a concept for orbiting electrons based on probability and other quantum laws of particle behaviour. It proposes electrons exist in "probability clouds" and there is only a defined probability (less than one) that an electron at a specified energy level is in any place at any time. The probabilities are controlled by various quantum laws which are connected to photon energy and wavelength. But there seems no reason why this general concept for the dynamic physics of electrons should not also apply to protons. In fact it seems obvious to me that it must apply to both charged particles in the same way. Equivalence or at least similarity in general atomic behaviour between electrons and protons must surely apply but scaled of course for energy differences.

So it mean that whatever the electron can do, the proton can do and perhaps do it just as well but with much greater energy involved? The proton orbits must also be compatible with "atomic orbital theory" based on wavelengths. In this theory all particles have a wavelength (wave particle duality) based on their energy content or mass and this wavelength determines "orbital" behaviour. So protons must have a much smaller orbital radius compared to electrons. Does this concept automatically prove the potential existence of proton light or proton photons! Perhaps not but it seems to add strong support to my proposal that a form of light or photons, can come from protons! Perhaps not complete proof but that will eventually come.

This proposal is a major break with conventional physics. The current atomic (macro) model has the protons fixed in the nucleus but I believe this is incorrect and propose that they are also in "orbit". I have not yet been able to find any previous or current proposal along these lines in my research into atomic physics or light theory. Perhaps my restricted access to the latest high level physics research is a reason for this but the complete absence of such an idea is more likely. It is so profound that it would be front page news if it was proposed by any recognised physicist. But as it has only been proposed by me, who knows or cares! That is about to change however!

Perhaps the next question is what are the dynamics of neutrons? Are they all held tightly together in a type of static solid nucleus as per the current model or are they also moving? They have no net charge and are overall electrically neutral but on a smaller scale they may have a polarised charge property. If a neutron is just a proton and an electron combined in some close "relationship" this would create neutrality but also polarisation. The neutron must also follow the wave/particle duality model and have a wavelength but without any charge it may not be dynamic. Also unlike the electron/proton pair, there is no obvious partner for any neutron to dance with, apart from perhaps other neutrons of different polarities, if available. Further study is required in this area so any interested physicist please take note.

So here is the new simplified planetary atomic model of the atom (the new Parkes atom).

A NEW ATOMIC MODEL

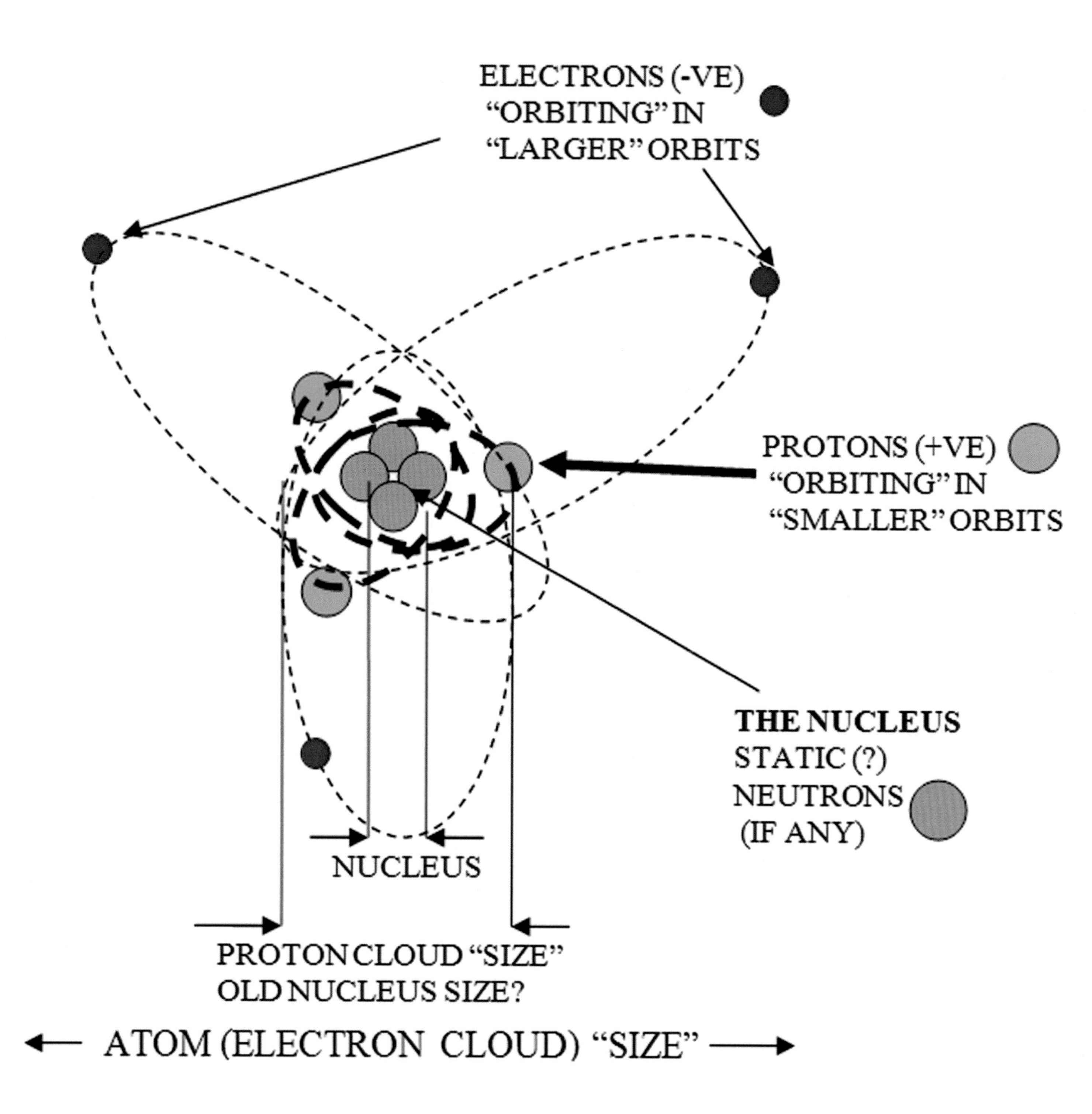

THE NEW "PLANETARY" ATOMIC MODEL
(NOT TO SCALE)

This concept of a new atomic model with orbiting protons may have implications for the original Rutherford scattering experiment that "proved" the nucleus had a hard centre, just like some chocolates I have eaten recently. However his experiment was very crude in terms of estimating the structure and size of the nucleus. It had no real way of determining any fine structure of the nucleus or of determining if protons were really fixed in the centre or just very close in small orbits. A solid nucleus was the most appropriate and simplest supposition. As orbiting protons have a very small orbital radius due to their very high energy and short wavelength, they would still seem like a hard core to Rutherford projectiles. Perhaps the real makeup of the nucleus will come to life in some bold new high energy scattering experiments using hadrons?

PROTON PHOTON (PP)

This is where a real jump in science is proposed. When I was developing my theory of gravity, light was central to it. Light pressure is what causes gravity through shadowing. But EP light doesn't have the energy or mass and hence the momentum needed to create the pressure differences required. It is too light so to speak. And if gravity was due to conventional light there would be too much pressure from the sun. I then used lateral thinking to consider alternatives. Is light really only related to one of the two charged atomic particles? If so is there any obvious reason for this? Is it possible that perhaps the proton could behave in a similar way to the electron in terms of atomic activity and hence be involved in light creation and capture?

To me there seemed to be no obvious reason why such proton activity can't happen. So I proposed that the positively charged proton can also create or capture light photons just as the negatively charged electron does. The detailed mechanisms are undefined but obviously a lot more energy may be involved in proton energy level jumps than for the electron. This is the basis of my proposal for heavy light and hence light gravity. The particular situations and types of atomic activity in which this process may occur are still unclear and are still under consideration but I am sure they will soon be identified. To be created, PP may of course require considerably more energy and heat than is provided by our suns activities.

Wow! This is new light indeed. Or to (mis)quote a famous space traveller,

Yes, there might be light out there......

......but not as we know it Jim, not as we know it.

The main initial difficulty I had with this bold new proposal was that perhaps light is not caused by electron energy level jumps but by a combination of electron and proton energy transitions. That is, the two particles work together to make light. However, it seems that current light theory is based solely on electron energy transitions. No allowance is made in "conventional" energy transition and light creation processes for any proton energy or other (quantum) properties. This is still an interesting area for further work but at this stage I have assumed that each atomic particle can act independently in energy transitions. Hence light creation and capture should follow a similar process for each of the two charged subatomic particles, only the energy amounts and hence the wavelengths would be different. Bold assumptions but let's proceed.

So now we have another type of light, proton light. I have called photons related to proton activity proton photons or PP, compared with EP for electron photons. It also stands for Parkes Photons but that is just by chance. Because of its source or origin from proton energy transitions, this new light is higher in energy and hence much shorter in wavelength by orders of magnitude compared to conventional light. And according to conventional physics, proton light should therefore have a much higher "mass" than electron light. So I have given this new type of PP light the name "heavy light", hence the title of this book.

WAVE-PARTICLE DUALITY

Light has schizophrenic characteristics. This is often called the duality property of light. It supposedly has both wave like and particle like characteristics or properties at the same time. Both properties have been applied to explain various behaviours but is seems that neither can fully explain all aspects of light. Current theory needs both the wave like behaviour as well as the quantised particle like behaviour of light, although most analytical theory seems to prefer the wave approach. PP light should therefore behave in the same way and hence be just as schizophrenic or perhaps suffer the same type of bipolar disorder as EP light. It will therefore behave in two ways or have two properties, that of a wave and a particle, at the same time.

The wave properties of light are primarily based on experiment. The particle properties are primarily based on theory of Plank and Einstein and others. The debate over which is correct or which better explains the various behaviours of light has continued to rage since then. How will PP light add to that debate? The whole proposal for PP light is based on a similar process for photon creation and a similar type of behaviour of the photons once created as for EP light. Therefore it should be expected that PP light has the same wave/particle duality behaviour as EP light. Therefore each PP should be like a flat disc of E/M energy with no dimensions in the direction of travel. This is the model I have used in discussions about PP light properties.

What other properties of light may also be relevant to proving the existence of PP light and especially helping to devise ways to create, control and make use of it. Properties such as refraction, which are explained by the wavelength dependence of light through a media or between two different media, may also apply to PP light. Reflection and dispersion properties must also apply to PP light. But how can these properties be used in experiments to control, capture and measure it? There may be some types of media that provide the necessary capabilities but finding them will be a challenge.

WAVELENGTHS OF HEAVY LIGHT

It is assumed that PP and EP are both photons and both have similar light properties. But if PP is associated with high energy proton activity which produces much shorter wavelengths, is it still covered by conventional light theory? And what wavelengths does PP light range over, what are the upper and lower wavelength limits? As described in my previous book, gravity is proposed as being caused by proton light pressure. PP light would have higher momentum and create higher pressure per photon than EP light, hence its relevance to gravity. But unknowns in the gravity derivation are the particular wavelength and the intensity of background gravity (PP) light. Reasonable assumptions could be made about these properties based on assuming some similarity to CMBR but is it enough to create the gravity we experience? I believe it is.

The major difference between EP and PP is assumed to be the energy levels and hence wavelengths of photons and therefore photon "disc size". It is well known that the proton is several orders of magnitude "heavier" than the electron so proton "orbits" should be several orders of magnitude smaller. Therefore proton light (PP) should be orders of magnitude "stronger" than electron light for a similar photon rate. And to continue the photon disc analogy, a proton photon must be several orders of magnitude "smaller" than an electron photon. Again not strictly scientific but let's stay with this general concept. Another significant difference is that while we can literally see EP with our eyes and detect it with electron based technology we can't see PP or detect it with current electron based technology. But its effects on any mass or body due to pressure created gravity can surely be felt.

The frequencies of typical PP photons would therefore be very high with very short wavelengths. From an order of magnitude analysis it is likely that typical PP have frequencies of order ~10 ^{24} Hz or greater, with corresponding wavelengths of ~10 $^{-16}$ M or shorter and hence energies of ~ 10 $^{-9}$ J or approximately 50 GeV or more. A simple analysis derives the relative energies and wavelengths that may be appropriate for EP and PP as set out in table 4.2 below. Of course these are first order estimates and much more work is required to support them.

PARAMETER	UNITS	PARTICLES		PHOTONS				
		ELECTRON	PROTON	GHZ	light	xrays	gamma rays	PP??
FREQUENCY	HZ	1.24E+20	2.27E+23	1.00E+09	1.00E+15	1.00E+18	1.00E+21	1.00E+24
WAVELENGTH	m	2.43E-12	1.32E-15	3.00E-01	3.00E-07	3.00E-10	3.00E-13	3.00E-16
ENERGY	Joules	8.19E-14	1.50E-10	6.63E-25	6.63E-19	6.63E-16	6.63E-13	6.63E-10
MASS	Kgm	9.10E-31	1.67E-27	7.37E-42	7.37E-36	7.37E-33	7.37E-30	7.37E-27

TABLE 4.2 – PROTON PHOTON (PP) PROPERTIES

While only one relatively low energy example for typical PP light is given there may be a broad spectrum above this figure and perhaps also below just as for EP light. But is there an upper limit for EP light and a lower limit for PP light? Can they overlap and if so what does this imply for atomic light creation and capture activities. More research again.

My gravity theory (TOLG) shows that gravity is based on proton light so this would suggest that gravity has a role to play within the atom or at what are normally called subatomic distances. It would seem from first principles that proton light would impact on protons as part of the creation and capture process just as electron light does with electrons but what about neutrons. Perhaps neutrons can't be involved directly in any light creation and capture but can they be involved in light collision and deflection. It would seem that from capture area discussions that EP are too big to "see" neutrons or protons and vice versa. But what about PP light, are neutrons affected by PP light in collisions and hence do they feel PP pressure? A centrifuge is often used to separate heavy isotopes using the mass difference based on different neutron count. Isotope differences can also be detected by weight or at least detected in a gravity scale device. So they must be subject to PP light pressure. More research again but it seems logical.

If gravity photons (PP) are created by protons the question must be asked, how are they absorbed again? This is another mystery and is probably related to the so called atomic binding process (strong/weak forces). This is perhaps why it is difficult to open up or smash an atom, especially the nucleus. To attack Hadrons requires extremely high energy or perhaps appropriate wavelength bosons! The strong nuclear force is very local and perhaps the reason is because the wavelength of any particle or boson required to trigger any action at this level needs to be very short. Very short wavelengths would then have no impact outside these distances.

THE SPEED OF PP LIGHT

One of the properties of light (EP light at least) that truly sets it aside from almost all other physical activities is its very high and constant speed. But what is the speed of new proton based PP light and what is the speed based on? Is it the same as electron light or is it different and if it is different is it slower or faster? And if there are any major differences what are the reasons for them and how do they impact on its behaviour?

The best way to consider the speed question is perhaps first to assume that all light is basically the same. After all, light is just made up of coupled E/M fields acting as photons. So assuming the same E/M theory applies to all E/M waves or light (photons), regardless of source and energy, then the same speed concepts should apply. That implies that the basic properties of the medium that determine the speed are the same for all types of light, even PP light. It seems therefore that PP light should behave in basically the same way as for EP light, apart from any possible wavelength dependencies of a media. For EP light the electric permittivity ε and the magnetic permeability μ determine the speed of E/M waves v, hence the speed of light c. This speed is given by;

$$v = c = 1 / \sqrt{(\varepsilon^2+\mu^2)}$$

But it is well known that for light in free space (a vacuum) the speed of light (EP at least) is not a function of wavelength. That is, all wavelengths of light travel at the same speed in a vacuum. So in a vacuum these properties that determine speed seem to be independent of wavelength. Therefore the speed of new light (PP) should in general be the same as EP light in free space. This is another major assumption but there is no obvious reason for rejecting it. Eventually experiments and research will verify (or disprove) this proposal.

It is important to note that the speed of gravity is currently proposed as being the speed of light "c" (EP light that is). It has been proposed in some alternate gravity theories that the speed of gravity is much faster than c. The basis of this proposal is the limited amount of drag due to latency in gravity that has been detected between fast moving bodies. If this was the case it would have a major impact on the speed of my new light which is the cause of gravity as set out in my original theory (TOLG). However analysis of most gravity situations based on G.R. which uses conventional light speed in the equations, confirms the speed of gravity is "c".

I have proposed that gravity is due to PP light pressure and hence gravity should work at PP light speed. So if gravity works at EP light speed, as currently proposed, then this aligns well with the assumption that the speed for PP light is the same as that for EP light, at least in free space. So it would seem logical the PP light speed in free space is "c".

The question then becomes does PP light speed also change in a media and if so how and why does it change? Is it slower or faster than EP light within some media? Within some media such as water, glass or even gas or plasma, it is well known the light speed determining parameters change as a simple function of wavelength. This effect is quite small but is sufficient to lead to a detectable difference in the speed of light as a function of wavelength. But do these parameters depend on any other properties of light? For example are they a function of the source of the light or the amount of energy in a photon or anything else? Longer wavelength light is supposedly slower than shorter wavelength light in most media. This causes refraction or light bending as observed when looking at a straw in a glass of water. This is also the basis for differential refraction or dispersion and the spectrum effect. But how much do they vary?

The next question is why do these properties vary by wavelength and what happens with very high energy, very short wavelength PP light. While any variation may be very small for the "normal" range of frequencies of electron EP light, it may become an important factor for much higher energy (higher frequency) PP light for some types of media. The variation as a function of wavelength may also become nonlinear. Or there may be two sets of values of these parameters for a media, one for electron based light and another for proton based light?

The difficulty with answering these questions is not just in determining the wavelength dependency of these parameters but in determining if some other atomic or quantum process applies for very high energy PP light. Within a media it seems that EP light interacts only with electrons. The light speed determining parameters that have been measured may only be based on electron behaviour and electron fields. But PP light interacts with protons (hadrons) and the field effects that drive PP light interaction may be different when protons are involved. This seems unlikely but is still an open question for determining the wavelength dependency and speed of PP in a given media. However at this stage I have assumed that similar rules apply for PP light. All that is required is to determine the speed and the parameters for PP light in a media.

EXISTING LIGHT THEORY

The basic properties of light are well known and reasonably well understood as explained by the current theory of light. This understanding is based on extensive experimental work as well as mathematical theory from optical experiments, through Maxwell's equations up to the latest quantum models developed by many excellent physicists. Light theory helped progress atomic theory and the relationship between atomic electron energy states and light was born. This relationship created another scientific revolution, quantum theory, applying to both light and atomic physics. The two areas of light (EP) and atomic physics have been closely linked ever since. But how does PP light fit in with existing light and quantum theory ?

Maxwell's equations were originally based on electron related E/M field theory. But they don't seem to make any special distinction as to what causes the fields. So it can reasonably be assumed that the equations apply to both EP and PP although the concept of proton fields is a new one. The electromagnetic relations in Maxwell's equations use properties such as permittivity and permeability? Are they the same for proton activity as for electron activity and hence the same for both types of photons. If they are, then the same equations should apply. Therefore the light speed derivations for EP light theory should be applicable to PP light. But this is a very challenging issue and requires much more study. Where is the new Maxwell?

Planks theory of light quantisation was based on EP but seems to be a general concept which makes no distinction as to light source, energy levels or other such factors. It therefore seems likely that Planks theory applies to PP, just as it does to EP. And it would seem that the Plank constant used in the equation for the energy of electron photons or EP light could also be used for proton photons or PP light. Einstein's work on the photoelectric effect would also seem to be scalable. Therefore there is every reason to believe that the same concepts apply to PP light from proton energy level jumps. The capture or release of quantised energy in proton energy jumps should be similar to conventional electron energy jumps. I have used this similarity in all my derivations of energy for PP light. However this is another interesting area for further study.

Einstein's work on relativity and light speed was for EP light. Does it also apply to PP light? If the speed of both types of light are the same in a vacuum and if there is a potential overlap in the wavelengths of the two types of light, then Special Relativity should also apply. In fact I have used the famous equation relating energy and mass that Einstein derived based on EP light, in my calculations for PP light properties. However, the interpretation of what PP light means for General Relativity is a challenge that I don't have the skills or time to address apart from believing that PP light causes gravity. More work for some bright young physics students.

The only significant difference between the two types of light seems to be in the energy levels and wavelengths. PP light has much higher energies and hence much shorter wavelengths than EP light. Or in terms of my photon disc model, PP discs are generally very much smaller than EP discs and the capture process will need to take this size into account. It is well known that EP light covers a broad spectrum of wavelengths and hence spread of energy levels over many orders of magnitude. It would therefore seem plausible that PP light would also have a similar broad spread of energy levels and wavelengths but from a much higher starting point. Is it possible that these two spectra overlap in a limited way around the wavelength of Gamma rays? All other properties seem to be the same so the overall behaviour should be the same I believe.

But the question remains how can PP light to be "observed". Current light technology is electron based and this technology may not be able to detect extremely short wavelength PP light. Gamma rays are currently detected by electron technology such as Geiger counters but this may be an extreme exception. Therefore, some form of new proton based technology needs to be developed to enable PP to be captured and measured. The properties of PP may then be measured and assumptions of similarity or should I say equivalence may then be evaluated. Until then I will continue to apply light physics developed for EP, for PP light calculations.

Imagine a mirror that reflects PP light, reduces pressure and hence reduces gravity behind it!

MASS - ENERGY, MOMENTUM,PRESSURE

From the work of Maxwell, Plank and Einstein, light in a quantised form of photons has quantised energy and an equivalent mass and therefore momentum. Each photon has a mass directly related to its wavelength of frequency. An estimate of wavelength and hence energy enables a photon mass to be derived and together with the speed, c, this enables momentum to be derived per photon. To derive light pressure, this momentum together with an estimate of PP light intensity is required. However, the rate of PP light that causes gravity is a great unknown at present. I have coined a term called Cosmic Gravity Background Radiation (CGBR) to describe gravity radiation. It may be possible to approximate it by assuming it is a similar type of universal background radiation to CMBR and using CMBR intensity measurements.

A typical spherical body of say 1 K.gm. with average density materials has a frontal area of about approximately 0.01 sq. m. Therefore the pressure required to support this mass must be 100 K.gm. per sq. metre. This amount of pressure would be the CGBR pressure difference that causes the gravity effects on the body. This figure together with estimates of PP (CGBR) wavelength or energy would enable an estimate of CGBR intensity necessary to create this gravity pressure, to be determined. Allowance needs to be made for angle of incidence and the angle of reflection of photons which will each reduce the pressure by a factor of less than 2 but together less than an order of magnitude. The other factor which impacts on light pressure from light striking a surface is related to the amount of reflection versus absorption. Again this involves at most a factor of two. However the current missing link is the necessary TOLG mathematics to derive a shadow causing pressure difference. When the complete TOLG theory is developed this could then be used to produce an estimated background CGBR rate.

But how can CGBR be detected and measured? Of course it may be detected and "measured" by gravity but this is the conundrum. Using gravity assumes the theory is correct so can't be used to prove it. So what other means of measurement could there be? This is the big question.

OTHER LIGHT BEHAVIOUR

Perhaps another way to investigate this new light is to examine other properties relative to light we know. The basic properties and "behaviour" of EP light that can be readily observed include the following:

Transmission in a straight line within any "transparent" media;
Reflection at certain media interfaces under certain conditions;
Refraction and/or dispersion at media interfaces under certain conditions.

It can reasonably be assumed that these properties apply to all wavelengths of light given the appropriate conditions. All light must travel in a straight line (geodesic?) unless deflected by some material or "force". In free space this implies that PP will follow the same transmission laws that apply to EP. And as my theory of gravity states that PP pressure causes gravity, then gravity PP can't deflect or bend other PP. It is well known that E/M fields can influence EP light behaviour. But can E/M fields cause deflection or behaviour changes to PP? If so what types and magnitudes are required? Is this a path to detection, measurement and control?

For reflection, it is well known that EP light is easily reflected by most surfaces. Some surfaces produce highly regular EP reflection that enables their use as mirrors but in general reflection is chaotic. But as PP light would seem to be able to penetrate almost all types of conventional atomic media very easily, how can reflection occur? What types of special material or even field effects could cause PP reflection. Perhaps some form of plasma or non-normal atomic states such as Bose/Einstein condensates could do the job. If such material is found it could possibly pave the way to controlling or at least managing gravity.

Light, EP light that is, can be bent as it traverses the interface between certain transparent media. What happens when PP light traverses certain media interfaces? Perhaps some very heavy metals or material doped with heavy metal isotopes may provide a measureable refractive index effect for PP and this should be followed up. Can broad spectrum PP light be broken down into a spectrum of wavelengths by some form of dispersion following refraction and if so how. These are all very interesting questions that will have to go unanswered for now.

PROVE IT

So now to the finale, how to prove this new light exists. The big problem is how can PP be created, detected and measured in a controlled and repeatable way? Of course it is "detected" and "measured" by gravity but this is a key part of the theory and hence can't be used in any proof. What other means of controlled, repeatable experimental creation, capture and measurement of PP could there be? This has been the focus of much of my work since I first proposed the idea of heavy light for my theory of light gravity. Do I have a sound scientific proof? No or at least not yet! Do I have an extensive or even limited mathematical analysis? Again no, or at least not yet! But I do have some sound scientifically based reasoning behind my idea as presented above as well as some practical research proposals for further consideration. Light from proton activity, called proton photons or PP (or in the case of gravity light which is proposed as being limited bandwidth PP called CGBR) is hard to "see". These photons have very high energy and very short wavelength. PP would have a very high degree of atomic penetration as most of an atom is free space. Electrons would not be able to capture PP light, they are the wrong "size" or they have the wrong "capture area". Detection by existing electron based light technologies would therefore be problematic. Perhaps PP may be detectable using electron technology such as a Geiger counter in special cases due to some secondary effects. It is possible that PP collision with a nucleus in limited cases causes some other transition which produces EP. Could a PP radiometer be made by shielding it in one direction? This would be a challenge as PP is so penetrative and it would also require very thick plates to capture enough PP.

It could be possible to create PP from controlled proton energy "transitions" just like EP are created from controlled electron activity. It may be possible to create PP from proton "collisions" using high energy proton colliders just like X-Ray EP are created from high energy electrons. If so, then perhaps this may lead to an experiment for trying to create and capture or at least observe PP and measure their properties at high energy colliders. Unfortunately such experiments to create capture and "see" PP require the use of high energy particle accelerators and I don't just happen to have one of those. Perhaps PP may have unknowingly already been created in large quantities by proton collisions at CERN and other establishments. However, because PP had not been proposed or defined and are difficult to "see" in a conventional particle sense they have gone unnoticed or been incorrectly recorded as other events.

Another possible source of PP is from atomic nuclear decay or radioactivity. Can atomic reactions such as radioactivity or even fission be used in any way to "observe" PP? Is this the type of new technology required? Radioactivity or spontaneous emission from some elements may not really be spontaneous after all. It may in fact be triggered by random capture of high energy background PP. Perhaps radiation and atomic reactions by unstable high atomic mass elements or isotopes may provide an answer. Such elements may be used in a way to try and capture and count PP? An approach to measuring PP light could be based on using lightly shielded radioactive atoms such as U_{235} as detectors of artificially created PP by high energy proton collisions. A controllable source of PP may be established based on a known rate of proton collisions. This may produce a measureable increase in "spontaneous" radioactivity.

The biggest problem is the expected very high penetration of PP through even the heaviest of metals. Any collision, capture or detection would almost certainly be of very low order and perhaps even based on secondary effects and would be on a very small scale. Perhaps another Rutherford low probability event (one in many thousand) type experiment is required.

Some cosmic experiments or observations could also be thought up to investigate PP. Unusual gravity situations such as pulsars and rapidly orbiting pairs of massive stars, could be used to investigate PP and gravity together. This may be useful in the evaluation of different types of shadow gravity behaviours with latency and finite light/gravity speed delays. And what about the recent "discovery" of "gravity waves"? Are these just minor variations in the intensity of background PP (CGBR) that causes gravity? If so, they would presumably be more common. So why are gravity waves, or at least those detected by LIGO, at least 40 years apart? The current heavy water filled underground chamber (old mine) experiments set up to detect high energy (light speed?) high penetration particles such as neutrinos, may also be useful. These capture secondary effects of very low probability particles with high penetration. Perhaps these are already observing PP in an unknown secondary way or could be set up to do this.

So what other types of experiments can be created so that some form of controlled PP observation, capture and measurement can be carried out? This situation is reminiscent of the search for the neutron. It was difficult to see a neutron using old technology such as cloud chambers with E/M fields. It required a new approach and new technology. But what new technology will enable us to readily create and see PP? Existing research in the field of high energy particle physics should be able to shed some light in this area. I look forward to someone in the right place with access to the right equipment taking up the challenge. Anyway let the search begin for any method of making, capturing and detecting PP. Perhaps my new heavy light will eventually "see the light of day".

A HEAVY LIGHT SUMMARY

The main outcome of this chapter is my proposal that there is a new type of light called PP or heavy light. I say *new* not because it has just happened or recently been created but because I have recently discovered or at least proposed it. It has been there all the time and is all around us and like most light, we can't see it. But we can definitely feel it. Perhaps we don't realise what it does or means to us but it keeps our feet on the ground. This new form of light is just as important to our lives as all electron light, especially optical wavelength light which has enabled us to see our world. While it works in a different way it has enabled the creation of our solar system as well as our own creation and is necessary to help us keep our feet on the ground and continue on our perhaps somewhat rough road to survival.

Originally I wanted to call this chapter "A NEW THEORY OF LIGHT", but that didn't capture the importance of my new idea, hence I renamed it "A THEORY OF NEW LIGHT" just to correct the emphasis. While my idea is a new theory about light, it is also about a completely new type of light, not just a new theory about light as we know it. Perhaps instead of challenging existing light theory, it just adds a new dimension to it. In fact, this new type of light relies heavily on existing light theory to describe and explain it. As a result of this new theory of light or should I say, major addition to the existing theory of light, new opportunities and challenges for the ongoing study of light and the extension of light theory will arise.

Now all that is required is some form of scientific proof that this new light exists. I have been unable to do that but hopefully others will eventually find the theoretical and experimental proof required. Physicists currently studying what happens when high energy protons collide would be in an ideal position to initiate this research. The Higgs boson must be related to my theory in some way. This is a call to super CERN. Surely PP has been or can be created in this environment! They just need to be captured and measured in a controlled way. Are any physicists willing to take up this unorthodox challenge?

I hope that this chapter will raise interest and not just eyebrows in the scientific community. Some of my proposals, especially new light sources and my theories on a new atomic model are sure to raise the temperature but hopefully also the interest of physicist. They should trigger many reactions and hopefully some positive responses. I hope you will share that view, especially after reading this book. Perhaps you already have some exciting ideas on how to create, capture and use this new light. I look forward to any comments from anywhere but especially from the scientific community. I hope they are mostly positive but in any case constructive, either to support and prove my ideas or to scientifically disprove them.

CHAPTER 5 – SPECTRA AND RED SHIFT

There are two fascinating and important properties of light that control much of its behaviour. They are closely related and a sound understanding of both is fundamental to the analysis of any property of light. The first one is that light, like the famous cloak, is made up of many colours. The other is that light is made up of many particles (photons) and each one has a fixed (quantised) energy level and a specific wavelength or colour. But how do these two properties influence the behaviour of light and what is their role in the spectral processes? In any case these two properties are critical to spectral and red shift analysis.

These two surprising facts about light took some discovering. While such proposals had been made many years ago by some scientists, others at the time and many since had alternate ones that gained more support. The true picture required new advances in technology and in theoretical physics before it could be shown. It was only recently that the spectral properties of light and what it implies for atomic physics were more fully appreciated, understood and scientifically explained. The quantised or discrete property of light also wasn't scientifically formalised until relatively recently. At the time there were serious questions about this new theory of a quantised form of energy. Even now there are still many unanswered questions about these properties of light and how they are controlled by or related to atomic activity.

Newton played an early key role in these areas of light. His experiment using a simple glass prism to make a rainbow may not seem so surprising today but was at the time. These days it is a common science experiment at schools. But what he thought about his observations and the theory he developed was very challenging. He correctly proposed that white light is not "pure" light but is really a mixture of colours. Many scientists at the time believed that white light was the pure form of light and that colour was an imperfection caused by reflection or refraction. Newton realised that white light is made up of a mix of colours and these "components" are the more fundamental building blocks. He thought that his prism experiment clearly proved his theory was correct. But it was strongly challenged by many leading scientists at the time and long afterwards. Newton didn't get the recognition he deserved until much later.

If Newton had observed light through his prism from other less homogeneous sources and seen the spectrum in more detail he may have noticed other fascinating spectral phenomenon. He may have seen discrete line spectrum or breaks in an apparently continuous spectrum and found that light is made up of many (infinite?) lines of individual colours. Each very narrow line provides information on the substance (energy level and atomic structure) that creates it. If this spectral effect had been observed and combined with his corpuscular theory of light he may even have arrived at a basic form of quantum theory long before it finally surfaced hundreds of years later. But the technology wasn't there and he would have been too far ahead of his time. As it was, some of his basic ideas on light were so bold at the time that they had already put him in a difficult position with some of his colleagues, especially Hooke. With such a radical proposal he may well have been considered a mad scientist and completely ostracised.

The evolution of the theory of light struggled for a long time after Newton. It was driven mostly by the myriad of light interference experiments and the development of mechanical wave theories to explain them. The real discovery and proof of many of the current aspects of light theory had to await major developments in physics and mathematics. The detailed study of light, including line spectra, red shift and of course the resulting big bang theory, started only recently. It began slowly but expanded significantly with the discoveries of electromagnetic connections, quantisation, relativity implications and electron orbits. It eventually became a major field of physics and today plays a fundamental role in so many other fields. But the story of light spectra and light quanta is not yet over. There is still much more to discover and learn. Light theory in these areas may be due for a major overhaul.

WAVES, PARTICLES AND SPECTRAL LINES

So light is made up of very many very small bundles of energy or "particles". Each individual particle or photon of light is mono-chromatic. That means it has a single frequency and fixed energy level. By that I mean that each particle of light has a specific wavelength or "size" and is made up of or contains a specific amount or quantum of energy. The wavelength and energy quanta are related by a simple formula or law.

But light also behaves like a wave. There are numerous observations and mathematical theories that support wave behaviour. It is well known that for other types of waves, such as sound waves, a media is required for the wave to travel through. Of course the media which carries a wave doesn't have any net movement, only the wave effect which travels through the media. There is no general forward motion of any physical entity, just the disturbance and the energy of the wave as it travels along. Light "waves" or particles of light from a particular source travel along together at the same very high speed through space or a vacuum. But for light there is no need for a medium or "ether" for the waves to travel through.

The fact that light apparently has a continuous spectrum but is made up of discrete photons of energy each with a discrete wavelength is extremely interesting. This apparently continuous spectrum of light has been studied at length and a special branch of physics called spectral analysis has emerged. Spectral analysis is now well advanced and used in many aspects of physics, especially atomic physics. And while visible light is often the main focus of this discussion, all ideas or theories in this book apply in principle to all wavelengths of light. So once again I am using the term light to represent all wavelengths of electromagnetic radiation not just visible light. This is done for simplicity and hopefully should not cause any confusion. The same theories or basic processes and rules apply in all cases although the media as well as photon wavelength or energy factors may play an important role.

Light has other important properties including luminance and intensity. Luminance is usually a measure of the total light power emitted by a source. Intensity is usually a measure of the light power being received at a location or by a device. Chrominance is another property which is a measure of the spectral spread or mix of wavelengths of light from a source. If the light is made up of a broad mix of photons with many different wavelengths it is usually called white or pink light depending on the spread or mix. This light, also called pan-chromatic, will show up as a broad band of wavelengths on a spectrum. But it is still made up of discrete lines due to quantisation. And these lines may supposedly overlap due to line width spread.

If all the photons (light) from a source are of the same wavelength (colour) the light is called mono-chromatic. It will show up as a single line of a particular wavelength in a spectrum. However even for a monochromatic source there may be a distribution of wavelengths and energies causing a spread in the width of a line. The shape of this distribution is possibly similar to a bell shaped curve and would have an effective 3dB or half power bandwidth. The likely causes of line spread raise interesting questions. Current theory is that it is caused by atomic thermal movement but I have my doubts about this explanation. This is related to the question of the applicability of Doppler shift to light and is discussed later.

Light with a flat chrominance spectrum or distribution of colours is generally detected as white by broad spectrum detectors such as our eye. Such detectors are unable to see small amounts of colour bias and all colours are seen to be equally present. But it is possible to filter out specific colours using material that absorbs particular wavelengths or to separate them out using devices that produce a dispersive effect or differential refraction such as a prism or grating. A search was underway to find out how light was related to other areas of physics, especially atomic activity and in particular how it was created and captured at the atomic level. Scientists started using optical devices to try and discover more about the composition and properties of light.

This led to the discovery of line spectra by Fraunhofer, another famous physicist. The spectrum of the suns "white light" could be examined in more detail because of the great developments in scientific technology. When the supposedly continuous spectrum of sun light, obtained through a prism, was very closely examined it was found to have gaps in it or to be more precise the spectrum had dark lines across it. Some of the supposedly continuous spectrum of light from the "white hot" sun was missing. The lines were found to occur at specific colours or wavelengths. Physicists then tried to find out what caused these lines of missing light (now called Fraunhofer lines) and why they were only at specific wavelengths?

Spectral observations of controlled light sources in controlled environments also displayed similar patterns. In some cases similar dark lines were noticed from a continuous man made white light spectrum. It was believed that absorption lines or energy gaps in white light were caused by the gas which surrounded the source then absorbing some of the otherwise continuous spectrum of light. These are now called absorption spectral lines. In other situations from some experimental light sources only lines of a specific colour or wavelength were observed, now called emission lines. In the case of the sun, the absorption lines were soon determined to align with emission lines of particular well known elements such as Hydrogen. It was then realised that atoms in the gas surrounding the sun were absorbing light at these wavelengths. Some new lines in the suns spectrum were then identified and that is how the element Helium, a word which comes from the Greek word for sun (Helios), was discovered.

Then scientists started to use spectral analysis to examine atomic elements and determine the spectral identity of each type of element or atom. It was a bit like the current use of DNA to identify genetic lineage for persons of interest and other related bio identity questions. By examining the spectrum from a particular source the elements in its composition as well as surrounding gasses could be identified. If the source was originally assumed to be continuous white light, as is the case with the suns thermal based light, then any gaps or absorption lines were considered to be due to gases in the surrounding solar atmosphere or intervening gases or the earth's atmosphere.

Of course this is now all well-known physics but is worth reviewing the basics for the following discussion. To explain the phenomenon, constructed examples of what absorption spectrum (dark gaps) and emission spectrum (bright lines) may look like are shown in figure 5.1.

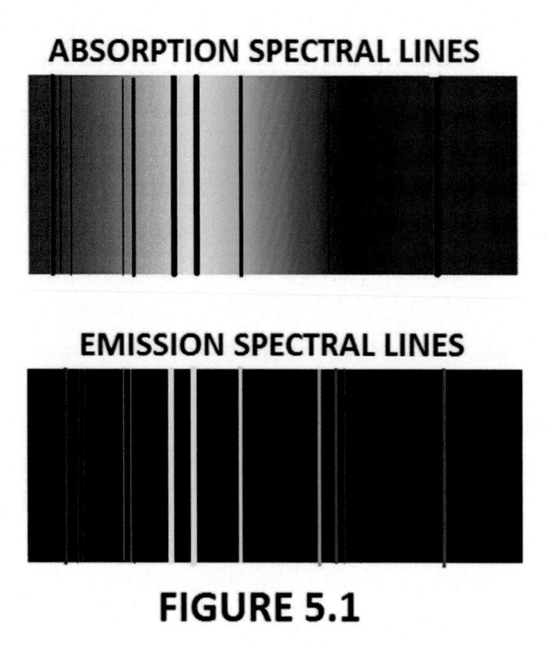

FIGURE 5.1

Spectral analysis suddenly became a new tool of physics and was used to explore many of the questions being asked at the time. It also raised new ones. The search for the structure of the atom became a major area of spectral analysis. This eventually led to the connection between photon energy levels and wavelengths and electron "orbits", the foundation of quantum theory. Another fascinating new application was of course the study of light spectra from cosmic sources. At first cosmic spectroscopy was used to study and evaluate the composition and evolution of our sun. Soon spectroscopy was applied to light from more distant cosmic light sources. The science of nuclear synthesis was developed to determine the atomic chemistry of stars and the universe. But what was observed had and continues to have a profound impact on physics and especially cosmology.

SPECTRAL SHIFT SHOCK

Around this time Hubble and his colleagues were trying to establish a method for determining reliable cosmic distances. The original method was based on using specific light sources or "standard candles" as they became called. The luminance of these light sources is supposedly well regulated and only varies according to such parameters as frequency of pulsation or other measureable quantities. The Cepheid Variable, a particular type of star, was identified as a useful type of standard candle and it became a key distance estimator. Many were observed, measured and recorded and this data was used to make a distance map of the known universe. Hubble and many others were trying to answer the age old but at that time very topical questions about the size and age of the universe.

While Hubble was using "standard candles" as distance indicators he saw more than he had bargained for. What was noticed had a major impact on cosmic physics as well as many other areas of physics? During spectral studies of the stars, it was observed that there seemed to be differences in spectral patterns from distant sources compared with local reference sources such as our sun. In particular, distant cosmic sources didn't produce the line spectra for well-known atomic elements such as Hydrogen. Eventually it was noted that the spectral lines matched known sources if they were wavelength shifted or realigned. This observation, which is where the term spectral shift originated, turned cosmology on its head.

Using other distance estimates, Hubble noted that the amount of spectral shift was closely related to his estimate of the distance of the source. That is, the further away the source from earth, the greater the shift of light from it. Hubble then proposed that the observed shift was directly related to distance and a new theory and law (Hubble's Law) emerged. Hubble's law which directly relates the amount of shift to distance using a constant called Hubble's constant has been used successfully in many cosmic studies since. However Hubble's constant has gone through some variations although the value now seems to have settled down to a "reasonable range" of values. But there is still some uncertainty as to its accuracy and the laws usefulness.

Actually it wasn't the discovery of spectral shift that had such an impact but the interpretation of that discovery. What was inferred from shift, lead to one of the greatest developments in cosmic physics. It was proposed that spectral shift was not simply due to distance but was caused by movement. The interpretation of spectral shift due to movement was based on the well-known theory of frequency shift due to movement called Doppler Shift named after its discoverer. This theory is based on the well-understood phenomenon for audio frequency changes in longitudinal sound waves due to differential motion between source and detector. This interpretation started a revolution in cosmic observations and led to the search for any cosmic spectral shift. Spectral cosmology studies and physics rapidly expanded just like the universe it said to be doing.

BUT WHAT IS SPECTRAL SHIFT?

Most spectral shift studies, especially for relatively local bodies with reasonable intensities, are based on spectral line measurements. A reference source such as our sun or a suitable local source in the laboratory is used to provide a base line. The spectrum of the light received from a cosmic body being studied is then obtained. A specific set of spectral lines such as those from Hydrogen (e.g. Ballmer series) is then searched for and compared with the baseline. This pattern may be found in the cosmic spectrum but at a different location or wavelength to the reference source. The reference spectrum is then shifted in wavelength to align with the cosmic spectral patterns. Shifting the reference spectrum until it re-aligns involves a shift either toward the red end of the spectrum or the blue end. If toward the red end then this implies the observed light has undergone red shift and if toward the blue end it has undergone blue shift.

A constructed example of both red and blue shift is shown in figure 5.2.

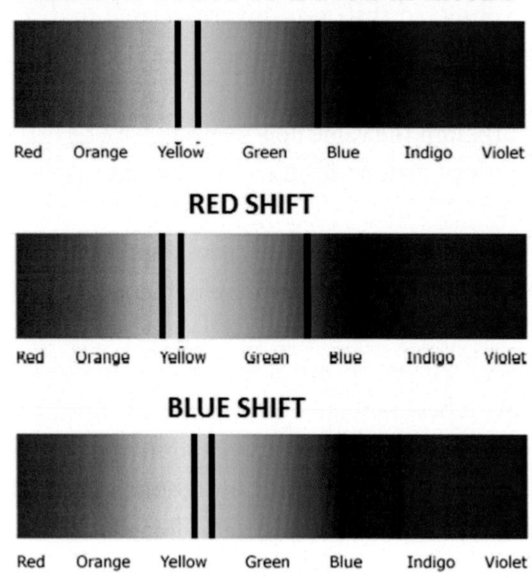

FIGURE 5.2

The amount of shift necessary to align the patterns is then measured. If the observed light has a longer wavelength for a known spectral pattern (emission/absorption line) compared to a reference source, then Z (the red shift metric) has a positive value and this is called red shift. If the observed light has a shorter wavelength for a known spectral effect then Z is negative and this is called blue shift. A red shift is due to an increase in wavelength and hence a reduction in energy of the light or photons coming from the distant source relative to a reference source. A blue shift is the opposite, a decrease in wavelength and hence an increase in energy of photons compared to the reference source.

Using Doppler theory, the radial velocity difference between the source and detector, v, is then given by the following formula. This is for lower non relativistic speeds where c is the speed of light. If this is positive then the two bodies are moving apart (red shift). If it is negative the two bodies are coming together (blue shift).

$$\mathbf{v = Z * c}$$

If the speed of either of the bodies involved approaches light speed then relativistic effects need to be taken into account and the light shift formulae need to be modified.

However for very low intensity, especially very distant sources where a complete spectrum may be difficult to obtain, spectral analysis is not always possible. In these cases an alternative method is used based on broad spectrum filters and light intensity measurements. Now I am not an expert on this type of spectral analysis so my interpretation may need to be modified but it goes something like this. The total light intensity from a source is first measured as a reference. The intensity of the light is then measured using a red filter and then a blue filter. These intensity measurements are then compared. If the intensity is greater with the red filter than with the blue filter then the source is supposedly red shifted and vice versa. However this method may have a fundamental flaw in the determinations of shift that I have called the sunset effect. I am not sure how this is allowed for if at all but the problem is discussed below.

Take our sun for example. It produces a broad spectrum of light, usually called white light. This spectrum is related to temperature and other factors. In the middle of the day with the sun overhead it is received here on earth as white light which has an almost flat distribution of energies across a broad spectrum of wavelengths, mostly in the visible light region. There is light energy outside the visible range but that is not important for this general discussion. If this light is examined through red and blue filters a similar amount of light intensity would be measured through each filter. The peak intensity of filtered light would be similar to the peak intensity of the unfiltered light at the filter wavelength. A simplification of this effect is shown in figure 5.3.

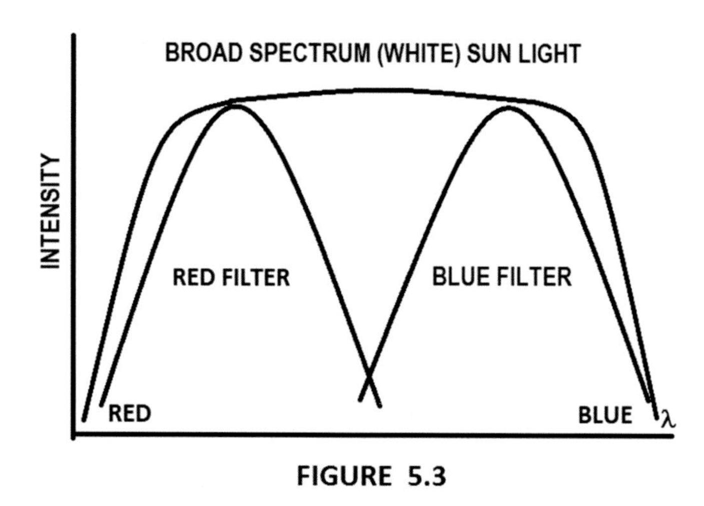

FIGURE 5.3

This process is used for light shift determination in most if not all very long distance cases. The light intensity is first measured without filters to get an average light intensity then it is measured using a red and then a blue filter. If the red filtered light intensity is greater than the blue filtered light intensity then red shift is assumed. On the other hand if blue filtered light is more intense then blue shift is assumed. In the case of our sun, as shown above, both filters would produce a similar intensity measure so no light shift would be determined as expected. Atmospheric restrictions which reduce the sun's harmful high energy UV rays (although perhaps not enough here in Australia to prevent sunburn and melanoma) would have no effect.

Now let's look at a sunset of our own sun. Late in the afternoon a red sun "sunset" is observed due to changes in the spectrum of light received. More red light than blue light is received and the overall intensity is lower. This is because sunlight has to travel through much more gas and dust at sunset. Red light is better able to penetrate the gas and dust while blue light is absorbed, deflected or scattered. Somehow lower energy light is less affected by dust and gas than higher energy light which seems the reverse of what would be expected. Perhaps this is because the wavelength or photon size of blue light is more closely related to dust particle size.

So when sunset light is measured through red and blue filters a difference in the intensity would be observed. Light through a red filter light would have higher intensity than the same light through a blue filter. This situation, for illustrative purposes only, is shown in figure 5.4.

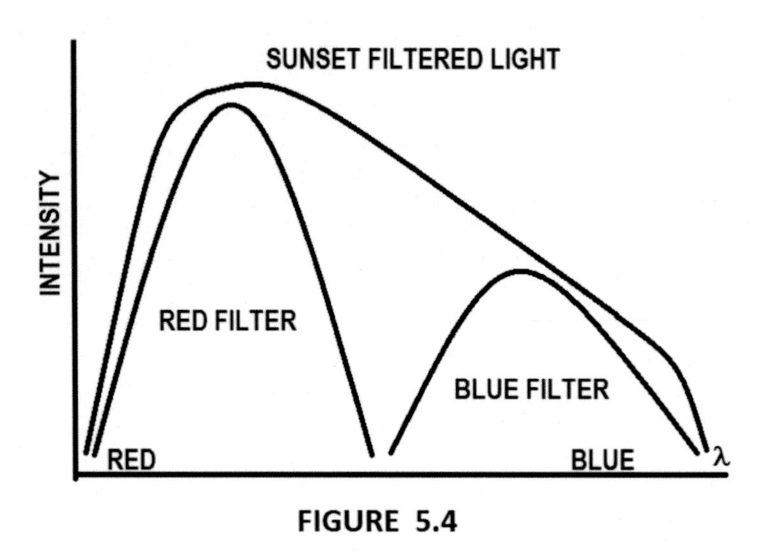

FIGURE 5.4

This difference in the intensities of the light measured through the two filters could be taken to be caused by red shift. But in this case that assumption is clearly wrong.

Now if the light from a distant cosmic source is red shifted this would imply all wavelengths suffer an overall shift toward the red end of the spectrum. This effect is shown in figure 5.5 below. The dotted line is the original or reference spectrum while the solid line is the new shifted spectrum.

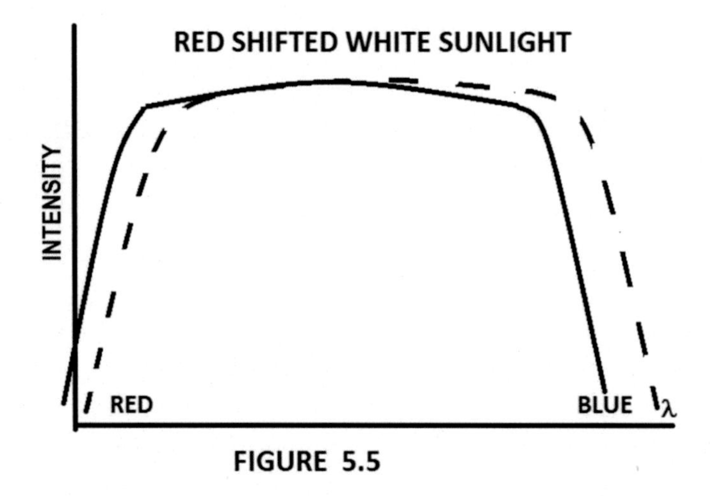

FIGURE 5.5

If filters are used to measure the intensity of this light, the red light intensity would not change but blue light intensity would be lower. This is shown in the drawing in figure 5.6 which has been slightly exaggerated to make the point. The change in relative intensities could be interpreted as red shift and the difference used to determine the amount of red shift or Z.

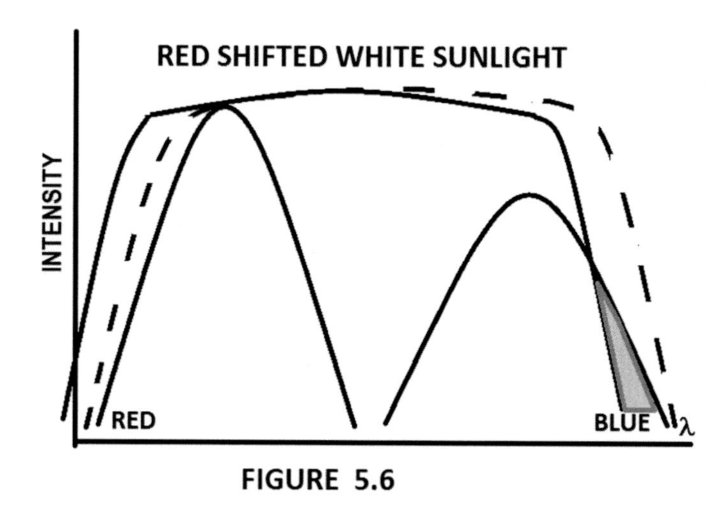

FIGURE 5.6

But is this really the only cause of intensity differences between red and blue filtered light? This method of measuring a distant cosmic light source so that red shift can be determined could involve a similar situation to the sunset effect if the distant source light has to travel through gas and dust. Is it possible that the sunset effect based on interstellar gas and dust may also cause observed "red shift"? This is shown in figure 5.7 below.

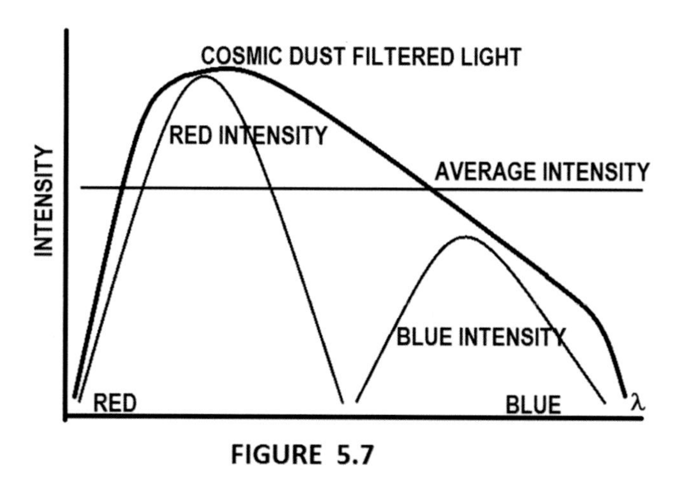

FIGURE 5.7

The intensity variation from this source could therefore be considered to have been caused by cosmic red shift. But perhaps it was just caused by a sunset type effect.

WHAT DOES RED SHIFT REALLY MEAN?

The initial interpretation that Hubble made was that red shift is distance related and it became a new form of "standard candle" for distance. Using the distance and the speed of light, it then became a model for the age of cosmic bodies and the age of the universe as a whole. The faintest and perhaps most distant sources of light with the longest red shift became the "edge" of the universe. Yes I know the cosmic model says there is no edge to the expanding universe. But it is a relevant concept when a finite age is proposed which implies a finite "size" cosmos.

Cosmic spectral shift was then proposed as being caused by differential movement between source and detector. Red shift was interpreted as movement away while blue shift was movement toward the observer. In particular the predominance of red shift of the most distant sources was taken to mean the universe was expanding. The cosmos suddenly became much more dynamic and interesting than previously thought. The expanding universe interpretation changed the understanding of the dynamics and just about everything else about the universe and shook the world of cosmic physics. This interpretation of red shift had very significant implications.

What does red shift mean to physics, especially cosmic physics? Well to answer the last question first, red shift is the basis of the current Big Bang theory and the related cosmic model of the expanding universe. The interpretation of large red shift at such long distances from earth was that the universe was expanding. Then some bright spark reversed the universe clock in a thought experiment to "see" what may have happened at earlier times. Assuming continuous expansion and not some other form of dynamic movement such as oscillating movement, everything was collapsed back to a (single) infinitely small starting point when there was supposedly nothing in terms of space, time and matter or energy?

This is how the current so called singularity or starting point of the universe was derived. The clock was restarted, the singularity "exploded" and the standard Big Bang Cosmic Model (BBCM) of the origin of the universe "happened". The so called expanding universe in the BBCM also created a new type of shift called cosmological expansion shift. This was used to explain all large scale very distant overall red shift observations. The interpretation of large scale red shift as expansion is critical to explaining the current BIG BANG based standard cosmic model of the universe. If large scale red shift is not due to movement, the BBCM cosmic model would collapse, not the universe itself, and there would be little left but cosmic (bull) dust.

But then other causes or explanations for expansion were identified. The expanding universe theory gained some acceptance when it was found to fit the new theory of General Relativity. That theory had initially proposed an expanding cosmos but Einstein, who originally believed in a static, infinite, flat and open universe, had already added a "fudge" factor to remove this expansion. He later said it was his "biggest mistake". But did Einstein really make a mistake in G.R. and how well founded is the BBCM?

One variation of the BBCM proposes that everything expands rapidly until it thins out to nothing or at least a lifeless cold thin isotropic cloud of dust and gas. Various alternatives from a slowing down of the expansion to a collapsing universe, including the famous continually oscillating one, have also been proposed. But as a true Copernican I can't believe we are just fortunate enough to be alive in the right place at the right time in any of these dynamic models. Hence I have great difficulty accepting anything other than a simple cosmic model of a flat, open, infinite, static and uniform (on a large scale) universe. Not very scientific but that is my current position.

As I said in my first book on gravity, something called shi(f)t happens. There can be no challenge to that fact. A number of extensive studies or surveys of cosmic light shift have been undertaken to map the amount of shift across the universe. They include the large scale American Sloane Project called the Sloan Digital Sky Survey or SDSS and the Australian 2DF Project carried out at the Anglo-Australian Observatory at Siding Springs near Parkes in Australia. These are available for studying properties, locations and amounts of cosmic red shift for research and to evaluate new and old cosmic theories. But I haven't seen them yet!

All these observations are based on extensive and respectable science and the data is very comprehensive and consistent. They indicate that distant sources and some relatively local sources show red shift while a limited number of local sources show blue shift, which in itself is a very interesting observation. There is considerable red shift in some cases, supposedly from very large and very distant light sources. On a large scale the distribution of spectral shift seems to be the same in all directions so the overall pattern seems to be isotropic. These observations can't be easily challenged but the interpretation of the data is surely open to alternative theories. But before alternative theories are discussed, known red shift causes need to be reviewed.

RED SHIFT CAUSES

One of the earliest questions I had as a budding physicist (unfortunately one that didn't bloom) was, why does red shift really happen? What is the true cause of this well observed and measured phenomenon? Is it just speed related (absolute speed or relative speed, whatever they mean), or is it due to other factors such as distance, acceleration, gravity, interstellar gas and dust, age or photon life time? Is it a simple linear combination or a complex function of some of these well-known causes working together? Or is it due to other physical properties or some unknown new physics? There are other possible physical causes of spectral shift type observations such as E/M fields and sunset effects where dust absorbs higher energy (blue) light more than lower energy (red) light. Special effects by transmission through transparent media such as deep water and thick glass where red light is absorbed more than blue light are well known but these are also not shift! Are there other causes of light intensity variations as a function of wavelength?

Current explanations for light shift are discussed below and summarised in table 5.1. It should be noted that in some situations more than one effect may come into play, even due to the same cause. It is also assumed in this analysis that all component effects are taken as acting separately if at all and can be linearly combined. These are reasonable assumptions for ongoing analysis but are still open questions. This is not a thorough analytical treatise on red shift but it is a scientifically based examination and provides a start for more detailed work. It is conducted primarily from a more practical point of view using thoroughly sound "mechanical" concepts based on light as discrete particles or photons. For a more rigorous explanation of red shift there are many more detailed explanations and mathematical analysis based on current theories. However some of these explanations and theories may still be open to question.

DOPLER SHIFT

This was the first and perhaps the most obvious explanation for optical shift. But what is Doppler shift? Well it really became a part of science when modes of transport such as trains started to get fast enough for the strange effects of moving audio sources to be noticed. The initial concept was developed by a physicist called Doppler who it was named after. He was investigating the physics behind the well-known observation of train sounds. As trains went faster railway crossings were made safer by the introduction of signal bells. Passengers could hear them go from a higher frequency ding, ding, ding to a lower frequency dong, dong, dong as they went past. Also, when a train sounds its bell or horn, a stationary observer notices that the frequency reduces as the train goes past. This affect applies to any movement with an audio source.

Doppler showed that this is simply a result of compression or expansion of longitudinal audio sound waves that are travelling at a fixed speed in stationary media (the air). The frequency shift is directly related to the relative speed difference between the source and detector and the speed of the longitudinal waves through a stationary media. A simple mathematical relationship between the change in frequency and the speed difference was derived. Doppler based frequency shift was then said to apply in all wave situations where there is relative movement between source and detector. Mach developed this concept further for audio waves when the source travels faster than sound. He defined a measure called Mach number which relates source speed to the speed of sound. Mach 1 is defined as when the source moves at the speed of sound.

It all seemed so simple to Hubble and most others. They applied Doppler shift theory to explain spectral shift of light which immediately related shift to movement. Spectral shift was believed to be caused by the Doppler like effect on "light waves" (based on the wave theory of light)) due to relative radial movement between source and detector. It was believed that if the source was moving away from the earth it causes expansion of the "light waves" and hence creates a longer wavelength causing red shift. It also seemed to explain blue shift which was also discovered from some cosmic light sources. Blue shift was said to be caused by the compression of "light waves" due to the source moving toward earth at high speed. Blue shift observations and the movement implications have led to many galaxy collision scenarios, albeit billions of years apart and billions of light years away but perhaps lasting for billions of years.

It was believed that all observed cosmic shift was due to relative radial movement between the source and earth. The source or detector or both could be moving in any direction but only the relative radial movement between the two caused detectable shift. So even if our sun is moving very rapidly (as would be expected by any Copernican) the earth would also be moving in the same direction. Therefore there would be no net radial speed difference and hence no red shift would be detected on earth. How convenient! Both red and blue shift have been detected equally in every direction from relatively local cosmic sources. But only red shift has been detected from very distant sources, again supposedly equal from all directions (isotropic).

A few simple calculations to derive the speed difference for some typical red shift measures found it to be phenomenal. Typical speeds derived from early red shifts were of the order of 1% of light speed. Larger ones, which imply a differential speed even greater than light speed, have supposedly since been detected. Current theory implies some bodies are moving away from each other at relative speeds greater than the speed of light but this supposedly does not imply that any particular body is moving at greater than light speed. But is this consistent with S.R or is it meaningful physics. This compares with the fastest man made spacecraft speed of 0.0001% of c and the estimated net speed of the earth (relative to what?) due to solar and galaxy orbits of about 0.001% of c. How do these relate to some extreme measures of Z and what does it imply?

The earth together with our solar system and even our galaxy must also be moving but at what speed and in which direction and in what relative sense? As a Copernican I believe our solar system and galaxy must be very typical in terms of position and place in the universe. (perhaps half way between the two extremes?). Therefore it must surely be moving at a very high speed, perhaps a reasonable fraction of light speed but let's assume a lower one for now. If the earth is moving at say 0.01% of light speed in some specific cosmic direction some amount of directional differential shift pattern from large scale cosmic sources should occur. It may be possible to detect such directional differential shift, depending on the minimum amount that can be detected using current best practise. But apparently no directional bias in cosmic red shift observations has been discovered. I am not sure if this is still the case.

The question then becomes what is the effect of the earth's movement in speed and direction? Are the observed red and blue or large scale red shifts only caused by movement of distant sources away from the earth? The earth must be moving at a very high speed due to so called cosmic expansion and surely only in one direction, not in all directions at once. Or is the speed of earth taken as zero in any direction or what? Does this interpretation imply that the earth is at the centre of the universe again or do all observation points in the universe see the same uniformity? If so how does this fit with the current view that there is no central point? What does it imply and how can it be proven. What if a bias in red shift is eventually discovered?

A uniform measure is what would be expected if the earth wasn't moving at high speed due to any expansion but just moving on a small scale in an infinite, homogeneous, isotropic, open, flat, static on a large scale universe. Perhaps an MM type of experiment could be conducted to find such movement but using spectroscopy. It would require the detection of many cases of shift over a long period covering many different directions. This experiment should surely be within current capabilities. A directional bias of wavelength shift has supposedly been detected in CMBR but in which cosmic direction and does it change over time? And is this directional shift bias caused by the absolute or relative (to what) movement of the earth as proposed above or something else?

An interesting analysis would be to determine the distance at which blue shift "drops off" and beyond which only red shift occurs. Beyond this distance (sphere around the earth?), blue shift would presumably still occur but would not be detected on earth. At this critical distance and beyond, perhaps any blue shift caused by local movement toward the earth even in high speed cases, is eventually overcome by so called cosmic expansion red shift. Another interesting analysis would be to include lateral movement of distant cosmic bodies to arrive at a net speed relative to the earth. This would allow for the fact that all local movements are in random directions but large scale movements are supposedly only radial and become more significant. This may provide a clearer picture of total net movement of cosmic dynamics.

But the real question is does "conventional Doppler theory" apply to light? Was this interpretation of red and blue shift correct? I don't believe Doppler theory applies to light and think this interpretation is incorrect for the following reasons. Firstly light is not a longitudinal wave. There are no longitudinal light waves like there are longitudinal sound waves. Light is made up of discrete photons that are in general not related to each other. Photons are two dimensional "discs" of energy that have no dimensions in the direction of travel. There can be no such thing as a "wavelet" (whatever it means) that has length or longitudinal dimensions in the direction of travel. Also light does not travel in a media (is the ether still dead?) and the speed of light is the same to all observers. Conventional Doppler theory simply doesn't fit the properties of light even though it may provide a scientifically sound explanation for spectral shift.

So what other theories based on movement could explain both red and blue shift? What type of photon wavelength or size change mechanism could be related to movement? What is really meant by differential movement between source and detector and what is meant by absolute movement which may be required to derive differential movement? Movement based light shift options including conventional Doppler shift are examined in a thought experiment.

MOVEMENT BASED SHIFT - A THOUGHT EXPERIMENT

Thought experiments are easy to create and cheaper than real experiments but suffer the problems of being purely artificial and perhaps never being experimentally verifiable. They can be repeated but the outcome is always imaginary and not in any sense real. "Results" are also open to more interpretation than real physical experiments. Thought experiments need to be backed up by sound theoretical analysis and hopefully by relevant actual physical experiments one day. But they do serve a useful purpose of identifying surprising outcomes and inconsistencies that need explaining as well as possible solutions to some questions.

There have been some very famous thought experiments, such as Einstein riding a light beam to work out aspects of relativity. He supposedly thought about viewing the town hall clock of a town in Switzerland as he rode away from it on a light beam. He imagined the clock would seem to stay still, stopped or frozen in time. This was than related to the relativity of time between two moving (inertial) reference bodies. It became a fundamental part of his Special Theory of Relativity. However, not all thought experiments have produced such spectacular results. The one about a falling person not being able to sense the difference between acceleration due to gravity or floating with no acceleration or a person in a lift using a light beam to measure movement was the basis for G.R. but these still raise some interesting causality questions.

In a similar way I thought of travelling with a photon of light to observe its properties as it changes. Of course the concept of "observing" the creation and behaviour of light without interfering with it may be problematic according to quantum theory but I believe it is still a useful thought experiment for discussion and analysis. The thought experiment is based on "riding on" or at least "observing" a photon as it is created at a distant (moving?) source and then staying with it as it travels, finally to be captured by another (moving?) body (earth?). In all cases the location, speed (whatever this means) and direction as well as the age of cosmic bodies and other factor**s** need to be taken into account.

The photon wavelength and creation effects play an important part in this thought experiment and are critical to this discussion so I have to be careful how these are considered. Unfortunately I am at the extremes of my knowledge on photon creation and am unsure about what parameters may be relevant in determining energy and direction of release but let's continue with this simple model.

One situation is for sources at a relatively local distance where both blue shift and red shift have been measured. They are assumed to be caused by movement according to Doppler effects. The second is for a greater distance source where only red shift has been detected. The interesting part is how the movement and speed of the source comes into play, if at all. The question arises, what is meant by the speed of the source? Is it the absolute speed of the body which could be defined as the speed of a body relative to the universe (at rest), whatever that means (something like Newtons or Mach's principle?). Or is it just some other type of special speed due to BBCM whatever that means (a bit like "quantum special effects").

So imagine light photons being created long ago by a distant cosmic source just like our sun. Sounds like a movie promo but stay with me. And imagine that this source was very young but old enough to create conventional light with all the associated atomic spectral effects. Therefore this process is the same light creation process based on the standard quantum atomic electron energy level jump model we understand here on earth today. This is a major assumption but stay with it for now. Now assume that photons were created and sent off randomly and isotopically in all directions from this source, even from a single atom. All photons were emitted at light speed relative to the body which created them and go in certain directions determined by a range of factors that are not considered significant to this analysis.

Here are some basic assumptions about photon creation and eventual capture for red and blue shift situations for the thought experiment.

1. A photon is created by a single atomic electron energy level transition by an atom that is part of a distant cosmic light source and emitted toward earth (or where the earth will be at the appropriate time).

2. This creation process is assumed to be identical to that which occurs here on earth and which has been regularly observed and measured (another bold assumption!)

3. The source was moving below light speed away from earth or was moving in that direction relative to the position of the earth as the earth might not have existed then.

4. The photon supposedly undergoes shift at creation due to some movement of the source. Note that this shift is only based on the direction and speed of the source when the photon was created and not the current position, speed and distance and not with reference to earth which didn't exist then.

5. From special relativity the speed of this photon is the speed of light which is the same to all observers. Presumably this hasn't changed since the so called big bang creation of everything! (Another bold assumption!).

6. This photon travels all the way toward earth with this initial shift established at creation. No shift occurs during travel as a function of distance or time according to conventional Doppler theory. (Red shift in transit due to BBCM cosmic expansion is discussed later).

7. This photon is eventually captured on earth in a reverse process and undergoes further shift which would add to or subtract from the existing shift depending on the earths relative radial movement. If the earth (and the solar system) is moving away (relative to the movement of the original source body when the photon was created) this will increase red shift. If it is moving toward the source it will reduce red shift or vice versa for blue shift. It may even cancel out all shift as it does for light from our sun. This assumes the earth is not at the centre of the universe and hence "static" but is moving relative to the original source position (and relative to all sources?) when the photon is captured.

8. This photon is one of many received from this source over time and is assumed to have no "contact" or correlation with any other photons. This is a bold and complex assumption that perhaps challenges entanglement which suggests that in some situations photons may have some connection with others which somehow allows them to communicate at infinite speed. I don't believe in entanglement and it seems to me to be another "epicycle" used to explain some strange observation in BBCM and quantum theory. Nevertheless the assumption of photon independence will require much more consideration and analysis but will suffice for now.

9. The captured photon(s) is found to be wavelength shifted compared to its expected wavelength because the photon was believed to have been created by a recognisable source such as a Hydrogen atom and would originally have been created (before any type of shift was added!) at a known Hydrogen line (or some other known) wavelength.

Now if the photon has red shift when it is captured, when did that red shift occur? Did it all occur at the time of photon creation due to movement of the source or the creating atom/electron or was it partially created by movement of the detector? It seems that unless a photon changes wavelength and hence loses energy during transmission (which is excluded from Doppler theory) all red shift must have been established either at the time it was created or the time it was captured or a combination of both. But given that shift of both types has been observed on earth from all directions equally how can shift be influenced by the speed of the detector on earth? How can the earth be moving in all directions at the same time. Surely this implies any shift observed on earth must have been caused (at least primarily) by the speed and direction of the source. But this creates a static earth centred approach which is unrealistic or is this what BBCM requires? Cosmic physicists please help!

For Doppler shift to apply there needs to be some amount of relative radial speed difference between the two bodies. This must be derived from each bodies "absolute" speed relative to some arbitrary reference point. This speed "information" must be included in the energy and hence wavelength of the photon at the time of creation and at the time of capture. Each photon, including our photon, must have been given an amount of energy and hence wavelength according to the atomic process that created it. This would then somehow be modified according to its speed and direction relative to the direction of movement of the source body. Somehow this speed and direction information must be determined by the body/electron at the instant the photon is created. This situation may also change very dynamically so there must be a very quick mechanism! I hope you are still with me on this thought experiment.

In other words, before the energy level of the photon can be determined, the source body must determine its "absolute" speed and direction. If the photon is sent in the same direction that the source is moving then its energy level will be higher (blue shifted). If it is going in the opposite direction its energy will be lower (red shifted). Of course the energy difference will depend on its angle of trajectory relative to the absolute direction of the source. Some photons will not be Doppler shifted at creation if they are released orthogonal to the bodies movement but that is a minor detail in this general discussion. So when any photon, including our photon, is created it contains "information" on the ("absolute") speed and direction of the source body in terms of its energy level and hence wavelength. This is according to Doppler theory!

But how can any cosmic body "know" what its absolute speed and direction are and how can this information be imparted to each photon? Is this related to the principle behind Mach's bucket where the universe somehow influences everything? And how can each photon when created, have energy imparted on it or extracted from it? Does the creating electron somehow have different electron "orbitals" depending on its speed and direction of movement? And what happens to this energy? The conservation of energy law surely applies or is this law broken in this situation? If the law applies it would require that the energy difference be extracted from or imparted back to the moving source. The "front" of the body facing the direction of travel would have less energy while the rear would have more. This must be reflected in the distribution of energy and probably the temperature of the creating body. This would be a bit like directional evaporative cooling of a wet body due to movement though the air! Can this energy/thermal difference be observed on our moving sun or does energy flow through the body in the same direction of movement to rebalance the net energy and meet conservation requirements.

If all photons of the same type generally have the same energy and wavelength at creation, regardless of movement of the source, how can such a change be made at creation or shortly thereafter? How can this really happen? This whole Doppler shift concept is based on a source "knowing" its absolute speed at all times, and then using it and the release angle of a photon to establish the energy and hence wavelength of every photon. This all seems very dubious to me. At best this is complex "quantum" behaviour (which is able to explain everything!) while at worst it is "black magic". This whole situation is already alarmingly challenging and smells strongly of epicycles to me. It is certainly not practical or realistic physics. Perhaps quantum theory really has the answers and I await elucidation or is it hallucination.

Then there is the capture situation. The photon arrives somewhere (here on earth?) to be captured by some lucky cosmic physicist using some sophisticated shift detector (a prism and the eye). Somehow the speed (absolute) of this earth bound detector is then used to add or subtract energy from the photon just before or during capture. In fact all light shifts may be increased, reduced, created or even cancelled here on earth. But is the realistic, I don't believe so.

The Doppler solution seems to present many challenges. In particular, how does any moving body even a massive cosmic body on the edge of the universe, know what its speed and direction are in an absolute sense? It would require some form of universal reference frame like a field or yes even the ether. Could there still be some form of absolute energy matrix or field or ether throughout the universe. What is it made of, what are the energy density implications and many more such questions? Surely these concepts of a universal ether or field have already been exposed as fundamentally flawed or have they? In a way shift would allow a body to determine its absolute speed using light which seems to go against the concept of Special Relativity. This situation can't be possible, it is all too fanciful, quantum theories or not.

So what is really happening or more precisely what happed long ago to this happy little photon. And what about the myriad of earth bound "red shift" experiments based on photon creation and capture with gravity gradients using static and dynamic sources. These have been set up in unused lift well laboratories and have used movement to create conventional Doppler shift to overcome shift due to the earth's gravity. They have supposedly demonstrated, measured and explained shift due to Doppler movement that cancels shift due to gravity. But they seem problematic due to the many questions that I have raised above. More investigation required?

To me reality must be much simpler than this. Each electron and hence photon doesn't require or utilise any information about source speed, absolute or relative, at either the creation or capture processes. All photons created by the same atomic process must be created in the same way with the same amount of energy and hence wavelength regardless of any source movement of any type in any direction. And at capture, information on the movement of the receiving or capture atom or electron simply can't be relevant either. The whole Doppler approach to explaining red and blue shift seems to be wrong. So what is the process that creates red shift? Is it related to the process that increases the width of a supposedly single wavelength spectral line? Is it created by the source or the detector or by both or just by something in between?

The most feasible interpretation of light shift is that each photon somehow loses energy (red shift) or gains energy (blue shift) between creation and capture. It could occur due to something like a local source field effect just after creation or during travel to earth or just before detection on earth. However as both red and blue shift have been observed on earth it would seem unlikely that any earth based specific field or other localised effect would have any major impact. It seems more likely that shift occurs almost entirely during travel or at the source only. This seems to meet all the requirements of red shift. But what happens to the lost energy? Is it lost to the intermediate dust, gas or plasma or to field effects? Blue shift, if it is real, would require each photon to somehow gain energy. This could be gained near the source due to some localised field effect or during transmission. However it seems highly unlikely that photons could gain energy from interstellar dust and gas during transmission but who knows.

So it seems from this simple thought experiment that conventional Doppler theory can't apply to any type of light shift. Shift also can't be based on movement (absolute) of either the source or the receptor or any relative combination of their movements. Other more plausible causes of red shift are discussed below. But blue shift is still a mystery!

SPECIAL RELATIVITY LIGHT SHIFT

The theory of Special relativity has been used to explain some aspects of light shift. When two observers or frames of reference have relative movement in any direction (radial or other) toward or away from each other, Special relativity comes into play. In particular each observer sees that the others measuring rod is shorter and clock is slower. Therefore a moving body has a slower clock relative to a "stationary" observer (all relative!) and so any clock related activity on the other body will be slowed down. Atomic light creation, the frequency of which could be associated with an internal atomic "clock", will show a reduction in timing. So when two bodies move linearly relative to each other each observes that the light from the other has a lower frequency than expected or is red shifted. This effect would of course only produce red shift as all movement causes time dilation or clock slowing.

This effect has supposedly been experimentally confirmed but I am still sceptical. But what causes blue shift? This requires clocks to speed up which goes against S.R. so time dilation can't cause blue shift.

But a stationary observer (all relative again) would also see a moving body has shorter measuring rods. Therefore anything which requires a length measurement such as light wavelength would see a reduction. Therefore relative movement also causes blue shift which again could be derived using Lorenz contractions or can it? Both these interpretations also have strange energy conservation implications that need to be addressed.

Relative movement seems to cause both red and blue shift due to Special Relativity. And they seem to be opposite and perhaps equal so they would cancel each other out or have I missed something? Back to the drawing board perhaps. In any case S.R. shift requires movement at a significant fraction of light speed before any noticeable shift occurs so perhaps it is not significant for local sources. But what does it mean for those very distant sources that are supposedly receding or moving at greater than light speed relative to earth. Another challenge!

It should be remembered that S.R. states that nothing (no body, energy or information?) can travel faster than light speed so Z from S.R. can never exceed unity. Red shift Z can be derived from the speed or velocity difference using the modified spectral shift formula for relativistic speeds below the speed of light given by;

$$\mathbf{1 + Z = \gamma\,(1 + v/c)}$$

where;

γ is the Lorenz contraction factor,

v is the relative speed difference,

c is the speed of light.

GENERAL RELATIVITY (GRAVITY) LIGHT SHIFT

According to Einstein's G.R. theory for gravity, light is affected by gravity. I have my doubts but let's proceed. The theory proposes that any light leaving a body will lose energy (but not speed) and be red shifted as it moves away from the body. A conventional body slows down as it moves away from a mass and loses kinetic energy but it gains potential energy in the field. There is no net energy loss. Gravitational potential field energy may also be converted to speed energy by a body when it approaches a gravitational field. Normally kinetic energy is lost in terms of reduced speed but light speed is constant or is it? Light supposedly loses energy to the "gravitational field" as it moves away from the source of the field. Light energy can only be lost in terms of a reduction or increase in wavelength. So how does light gain or lose gravitational field energy and how is this explained for a photon?

According to G.R. theory, light from the sun is first red shifted as it leaves the sun up to the point where the gravitational fields balance. Then it is blue shifted by the gravitational field of earth until capture. Simple calculations could derive the order of net solar red shift from our sun due to G.R. It would be very small but has supposedly been detected. If this model applies for shift from distant sources the mass of the complete galaxy including our solar system may be involved in determining red shift from very distant sources. Then local gravitational field effects would somehow need to be catered for. The mass required to produce observed cosmic red shifts from very distant sources could be derived but observed large red shifts from distant light sources if due entirely due to G.R. must have been caused by very massive bodies. The problem is in allocating the effects to each cause! But according to my new theory of gravity (TOLG), light causes gravity and is not affected by it.

Therefore this shouldn't be detectable so what is really happening? I must follow up on these experiment to see if it really is detectable or if is swamped by other factors. Perhaps it can't be measured easily if at all.

GENERAL RELATIVITY EXPANSION (BBCM) LIGHT SHIFT

Another supposed cause of red shift, which is also related to G.R., is the big bang itself. The universe was supposedly created at time zero (but not at ground zero) and has supposedly been expanding ever since, possibly not linearly for various strange reasons. This expansion of the universe is supposedly not like an explosion (or is it?) but is due to the continuous expansion of space (and time?) which supposedly expands everything in it as well, especially distances.

The expansion of the universe supposedly expands the light which travels through it. This causes what is called cosmological red shift. The light "longitudinal waves" are of "standard size" when created but get longer over time. This length increase causes red shift or so they say. So cosmological red shift is not due to movement directly but could be indirectly related to a form of movement. Calculations from extreme red shifts indicate some very high speed "movements", even greater than light speed. And surely as any Copernican would surmise, cosmic light shift must be caused about equally by the source and receiver so our solar system and earth must also be moving at a phenomenal speed but relative to what?

The cosmological red shift expansion idea goes something like this. Light from very distant and very old sources was created long ago. As the universe has expanded over time, this "old" light has expanded or the wavelength has been continuously "stretched". Again this interpretation is based on longitudinal waves of light (whatever they are). But light doesn't have longitudinal dimensions so perhaps it suffers the same challenges as Doppler theory which supposedly also stretches waves but only at creation or detection.

However, with BBCM expansion, everything is supposedly stretched. Therefore all lateral dimensions must also be stretched. Therefore light photons which have lateral dimensions (in a flat disc sense) would undergo lateral expansion and hence become bigger resulting in wavelength expansion. In this explanation of course there can be no contraction or reduction in photon size and hence no blue shift, only redshift. So BBCM and cosmic expansion can't explain blue shift either.

But if there is an increase in the size of each photon so they have a longer wavelength and lower frequency they also have lower energy. Therefore somehow all photons are steadily losing energy due to BBCM stretching. Where does all this energy go? This theory seems to contradict the conservation of energy law. I am not sure how this fits in with the BBCM or its quantum version with various energy relations and dynamics. Quantum theories often contravene basic laws of physics such as conservation of energy but it all seems problematic to me. Is an expanding, ageing universe steadily losing energy or does it go somewhere such as into heating up interstellar gas and dust? And if so, how does energy loss due to expansion end up as increased cosmic dust energy? This all sounds very challenging but again I'm no expert.

Now when expansion or stretching of everything occurs there must be other side effects. If the universe is expanding and light has expanded what about everything else. Surely a measuring rod carried along with the light photon (not possible but stay with me) or a stationary one at the source or here on earth would have expanded or been stretched over time in the same way. The length of both would increase by the same amount. If everything is expanding then so too are our measuring rods and any other devices used to measure physical properties such as light here on earth or in every part of space. All physical systems (atoms/electron orbits) used to create light must also have expanded. So any photons created by atoms today will now be the same size (same energy/wavelength) as photons that were created long ago and have been stretched by BBCM expansion.

What this means is that recently created light that could be used as a red shift reference, must already be expanded and hence already be shifted? So any red shift created by an expansion of space would not be observable. How can it be otherwise! It would not be possible to observe such an expansion if you were part of it. If everything gets bigger together then who notices. Or does the expansion only apply to light or moving things and not to stationary things? All very strange to me or have I missed something again. As usual I await expert explanation. And again what happens to the energy. It would seem that cosmological expansion saps cosmic energy!

I can hear the experts saying it's all so obvious, you just don't understand quantum cosmic physics. Yes there is an element of truth in that but what is really happening in simple terms or is it no longer possible to explain cosmic physics in simple terms. Shades of quantum complexity here but the whole idea sounds a bit like "The Emperor's New Clothes" to me. Who is hiding or protecting what? As I don't support cosmological red shift or the whole BBCM theory this is not a problem for me. I am ready willing and able to say "the Emperor is naked"!

Interestingly the expansion theory based on BBCM is now supposedly the main cause of red shift at least from very distant cosmic bodies. It is "the standard model" for red shift. But there is still the question of local shift and especially how blue shift can be explained. If blue shift is not part of cosmic expansion theory is it still really due to local movement (Doppler effects)? But this is also challenged and so existing explanations in the standard model seem problematic!

LIGHT SHIFT DUE TO DISTANCE AND TIME

Either indirectly or directly, red shift is a function of distance, and yes, therefore of time. While red shift was once used as a cosmic distance measure or indicator, other cosmic effects have largely replaced it in that role. But it is undeniable that red shift, especially overall red shift seems to be a simple and linear function of distance, at least on a large scale. Hence with the constant speed of light, it is also a simple and linear function of the lifetime of the light in question. And as light from the so called edge of the universe took a long time to reach us, its source has also aged and moved further away. It would now be older by the amount of time estimated for the light to reach us. So if the shifted light we see from a distant source was emitted very early in the life of the universe (not long after the BIG BANG) everything was supposedly much closer together then. The universes hadn't expanded much then so the earth was very close (or would have been if we existed then) to this very young source of light when it was emitted.

So it should have only taken a short time (light years?) for this light to reach us if it was emitted when the universe was younger and hence smaller. But it has supposedly taken billions of light years to reach us so the universe must either be expanding at near light speed or it was emitted when the universe was already very large! If so then the universe was already very old when the light we have just received from this source was emitted. The distant source body must already have been very old when it created the light we measure (more than old enough to create light as we know it). This must therefore have an impact on the estimate for the age of the universe. Its current age could be at least double the time of light transmission. And the current size of the universe must now be very much larger than the estimate based on the distance the light has travelled as expansion has supposedly continued (even accelerated?) over this time or am wrong again? This seems to contradict the model but perhaps I am lost again.

TIRED LIGHT SHIFT

There is another nonstandard theory on the potential cause of red shift called the "tired light" theory. In this theory light supposedly just loses energy with age, a bit like me I suppose. It does not imply any differential movement between source and destination but the actual mechanism is not well explained. So how does it happen and in particular where does the lost energy go? This tired light theory has received little support from any recognised cosmic physicists and has no real scientific basis. However it may lead to other possibilities.

COSMIC "SUNSET" RED SHIFT

Perhaps large scale cosmic red shift is similar to what happens here on earth at sunset. Some light may be lost in transmission due to interstellar dust and gas. The loss is wavelength dependant with higher energy blue light losing more energy to gas and dust than lower energy red light. This loss is a simple function of distance and time. Therefore so called red shift may not be caused by movement or be related to Doppler shift. It may simply be caused by light passing through interstellar space. The further away the cosmic source is the more cosmic space pollution it has to travel through and hence the more blue light is impeded. This is what I call the cosmic sunset effect. The lost energy (blue light) is eventually captured by matter which heats up and becomes more active in cosmic processes. No movement of any kind is required. There is no expansion and no big bang so there goes BBCM. But again, what causes blue shift?

The explanations of red shift (and in one case blue shift) are compared in the table below. Most causes are problematic according to my analysis of various situations. My analysis is not based on complicated quantum theory or detailed mathematical BBCM theory. Perhaps it requires a review in these terms. However, my explanation for red shift is simple physics but how can it be proven. That is I what I intend to do perhaps with the help of other sceptical physicists. Cosmic sunset may explain red shift but can't explain blue shift. Is it a real phenomenon or is it a result of misguided interpretations of difficult measurements. If it is really movement based then how can a source know its absolute movement. Could it be caused by some local electromagnetic field effect at the source or by some local or intervening media or some other unusual gas or plasma surrounding the source which creates a blue light effect?

COMPARISON OF RED SHIFT CAUSES

CAUSE	(+) VES	(-) VES	Z (<< C)	COMMENTS
DOPPLER SHIFT	SIMPLE THEORY RED AND BLUE	NO LONG'L LIGHT WAVES	0.01 - 0.1	CAN'T APPLY TO PHOTONS REL./ABSOLUTE SPEED? REFERENCE FRAME?
S. R.	ACCEPTED THEORY RED SHIFT ONLY?	BLUE SHIFT?	~ 0	ONLY FOR RELATIVISTIC SPEEDS - SHIFTS BALANCE OUT?
G. R. GRAVITY	ACCEPTED THEORY RED SHIFT ONLY?	LIGHT GRAVITY NO BLUE SHIFT	NOT SPEED BASED	MASS AT LIGHT SPEED CAN'T CHANGE PHOTONS ACCORDING TO TOLG
BBCM G.R./EXP.	STANDARD MODEL LARGE RED SHIFT	ENERGY LOSS? NO BLUE SHIFT	SOME Z > 1 V > C	EVERYTHING EXPANDS? WHERE IS LOST ENERGY?
COSMIC SUNSET	SIMPLE PHYSICS. DISTANCE NOT SPEED	UNPROVEN NO BLUE SHIFT	NOT REAL SHIFT	ENERGY LOSS TO GAS/DUST SCIENTIFIC EXPLANATION DEATH OF BBCM
OTHER	SPEED BASED? RED/BLUE SHIFT	NEW PHYSICS?	SAME AS DOPPLER	COULD BE E/M RELATED REVIEW OF PHOTON MODEL, THEORIES, MEASUREMENTS

SUMMARY

Red and blue shift seem real enough but what really causes it? The proposed explanations of red shift and in one case blue shift have been analysed and discussed. There is no doubt that distance is an important factor, especially for red shift, but what about movement. There are so called local effects which seem to be based on motion as well as large scale effects which are said to be based on expansion of the universe? Cosmological expansion must expand everything and therefore would not be observable. Conventional Doppler stretching or compression of light "waves" can't apply to individual non longitudinal photons. Special and General Relativity and gravity don't cover all situations and seem problematic in any case. Most current model causes are not suitable and acceptable according to my analysis so what is really happening? And if cosmic expansion doesn't cause red shift then the current BIG BANG model collapses.

If shift is really movement based and only occurs at source and destination, then how can every individual body know its absolute movement and then use that in determining the energy of each separate photon depending on its direction? Perhaps there is some form of universal cosmic E/M field that provides such an absolute frame of reference, but again how profound. If it is not caused by movement of a body toward (a stationary) earth could it be caused by some local effect at the source? Perhaps there are other unknown explanations for red shift and also for blue shift. The single flat disc photon model may be incorrect and a photon may be a group of light particles with spacing related to wavelength and energy being replaced by energy density. This type of model changes everything but movement becomes a relevant factor.

One possible cause of red shift, apart from cosmic expansion and Doppler, is the cosmic sunset effect. This seems to be distance but not movement related. More blue light than red light simply gets caught up in the interstellar gas and dust. As a result light loses energy in the long transmission process. But how can blue shift be explained? Does it really happen or is it a quirk of the measurement methods just like the sunset problem in filter based measurements?

What about cosmic microwave background radiation (CMBR). The current interpretation is that it was caused by the Big Bang. It was supposedly once at a very much higher frequency (well above our suns white light temperature) and has been red shifted down to its current very low (relative) frequency. It has been suggested that it was "red shifted" by a Z factor of the order of 1000 since it was created due to overall cosmic expansion of the universe. But is CMBR just the background thermal radiation of the universe at the average temperature of the universe (about 4 degrees Kelvin?). A small directional difference in CMBR wavelength (red shift) has been measured but I am not sure of the amount or the direction. It seems to suggest either an uneven (non isotropic) universe or an additional red shift supposedly due to (absolute?) movement of the earth! This would seem to imply the earth is moving very rapidly but in which direction and compared to what? And how does Olber's Paradox fit with the theory of red shift? Perhaps it implies the sky should all be red not white. Or is this the real cause of CMBR?

So what is the real cause of shift? My analysis is not based on complicated quantum theory or detailed mathematical models but on simple sound rational logical physics. My explanation for large scale red shift is based on something like the simple sunset effect. But how can it be proven. And if an individual photon approach is adopted, it seems that any shift can't be movement based. But blue shift is still a mystery. Perhaps a complete objective review of all shift measurements and shift theories is required to resolve these questions. I would like to do that perhaps with the help of other sceptical physicists. In the meantime my search to obtain a useable copy of a comprehensive cosmic shift map of the universe (or any part) will continue.

CHAPTER 6 –REFRACTION AND DIFFRACTION

There are a number of properties of light, especially interference, that are related to its wave like behaviour. Other properties of light like reflection can readily be explained by particle theory. They are simply photons bouncing off a surface in an elastic collision where the angle of reflection equals the angle of incidence. But properties like interference, refraction and diffraction can only be easily explained by using light behaving as waves. This established the wave theory and nearly killed off the particle theory of light. It seemed obvious to early physicists that these more complex behaviours are caused by a wave like behaviour not particles. They involve light interfering to produce patterns that are similar to the way water waves behave when they combine or interfere with each other. Such wave like behaviour cannot easily be explained using light as particles or photons interfering with each other, or can it?

This chapter takes a new look at some of the interesting light behaviour such as refraction and interference from a particle or photon point of view. Current models propose outcomes based primarily on wave theory. Even a photon approach is based on a wavelet or wave packet for each photon. However in some cases these wave based models do not match actual observations and also seem inconsistent with what would be expected from a simple quantised light energy point of view. Some current theories and models of light interference are examined, contradictions exposed and reasons given as to why they may be incorrect. New interpretations of these situations are proposed based on a particle or photon explanation. They attempt to show how light particles may behave and interact in order to produce the observed outcomes.

This work may not be based on theoretical analysis or comprehensive research but it is based on actual experiments and observations using real test equipment as well as a scientific approach. While many of the fascinating behaviours discussed may not have been fully tested in experiments, most of the basic ones have. My ideas are based on interpreting these new observations as well as reinterpreting existing models and theories. I have used a more mechanical view of what may be happening to individual photons in each situation to derive new theories and new models of behaviour. Of course they are only theories proposed for further discussion but they seem to provide sound and rational ways to explain most light interference behaviour situations. The next step is to formalise these new ideas in more scientific terms using appropriate mathematical theory, based on particles and not wave theory.

This analysis uses the basic properties of photons described previously. Also it is assumed that all photons from a general source of light, especially white sunlight are unrelated or unconnected with each other in any way. This is based on an assumption that all photons are created in a random way. Therefore it is assumed that in general, there is no relationship or correlation between any two separate photons at creation. While there may be a minimum (possibly variable) time separation between multiple photons being created by a single atom, this factor is assumed to be insignificant. There may be some form of correlation or coherence imparted on pairs or groups of photons as a result of movement through a media or media interface or past a device such as a slit or edge. However these situations are identified and discussed and likely outcomes addressed in each case.

REFRACTION, DISPERSION AND THE SPECTRUM

Light can bend. This is called refraction. It is usually caused by light traversing different media or even gases with different refractive properties. Current light theory states that refraction may occur at an interface between two media. The details are contained in theories like Snell's Law and refraction mathematics based on wave theory. The difference in the speed of light in the two media is what supposedly causes refraction. It causes light which normally travels in a straight line (or is it a least action path or "geodesic"?) to deviate from a straight line under certain circumstances due to a change in the least action path dynamics. This effect is often demonstrated using a pencil in a glass of water that seems to bend at the interface between air and water when viewed from certain angles.

One key aspect of current theory is that the refractive index of a material is supposedly wavelength dependant. This implies that light bending at a media interface is wavelength dependant. While the speed of light through a media may be wavelength dependant, is refraction really wavelength dependant? If it is then what really happens at an interface between two media. If refraction (bending) is always a function of wavelength then dispersion would occur at the first interface. If this was to occur then a rainbow (spectral effect) would be produced for light at the first interface. This would then be observed in relevant experiments, but has it? The bent pencil always seems to be the same colour.

Perhaps the most significant outcome of differential refraction is dispersion. This involves breaking white light, or whatever the source light mixture is, down into its component colours. The spectral effect was well recorded and analysed by Newton. Since then, a full theory of light bending and spectral effects has been developed based on wave theory. These laws and theories have been used to explain many types of light behaviour involving optical devices. Two very popular types of optical device or prisms used to study these phenomena are the rhombic or parallel sided prism and the triangular or non-parallel sided prism. Prisms can be made of any transparent material but glass is usually used. I conducted my own experiments using various glass prisms.

The theory of refraction is analysed using various types of prism in the following discussion. Current theory states that differential light refraction which creates a spectrum or rainbow starts at the first interface between two media. It then increases at the second interface to produce greater dispersion and a broader rainbow. This current standard model interpretation of refraction and dispersion is shown in figure 6.1 below. It shows a single narrow beam of white light (shown in grey on the left hand side) entering a triangular prism at an appropriate angle. Each colour supposedly undergoes a different amount of refraction or bends at a different angle at the entry interface based on wavelength. It then follows a different (but straight line?) path through the prism. Further differential refraction and dispersion occurs at the second interface to produce a broader rainbow. The effect of refraction and dispersion has been exaggerated in the constructed drawings below to provide a clearer picture.

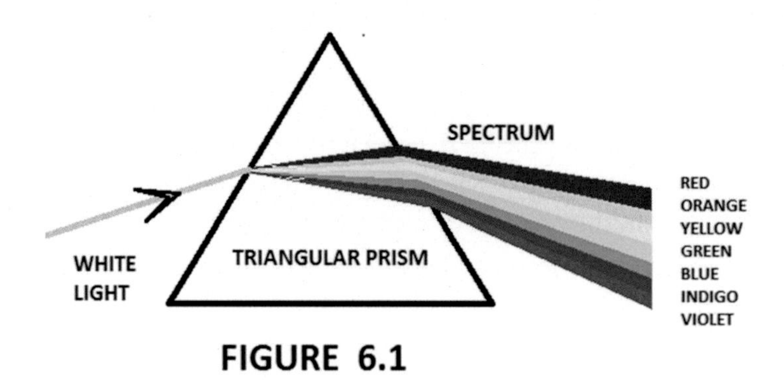

FIGURE 6.1

REFRACTION AND DISPERSION
CURRENT STANDARD MODEL

This is supposedly based on experimental observations but I am not sure. It may have been based on an incorrect interpretation of observations and theory! But if this interpretation of events is what really happens then it must also happen for a parallel sided prism as shown in figure 6.2 below. This shows a single narrow beam of white light (shown in grey on the left hand side) entering the prism at an appropriate angle. Refraction occurs at the first interface and each colour supposedly undergoes a different amount of refraction or bends at a different angle based on wavelength. Hence dispersion occurs at the first interface. Each colour then follows a different (but straight line?) path through the prism. Further refraction then occurs at the second interface but this would still produce a rainbow, perhaps still diverging? This is the standard model.

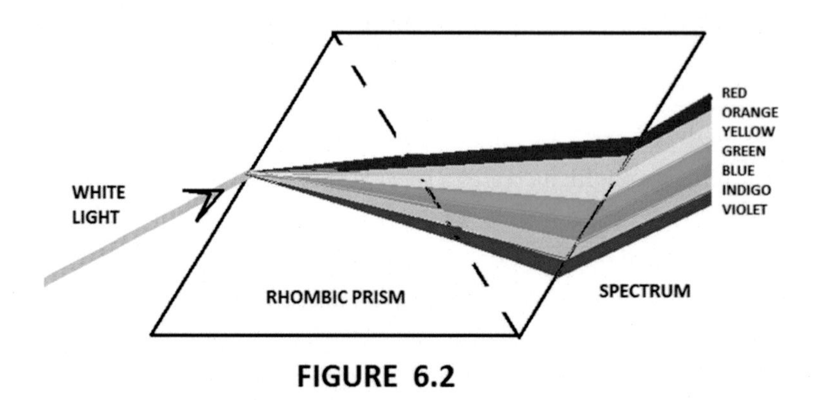

FIGURE 6.2

REFRACTION AND DISPERSION
BASED ON CURRENT STANDARD MODEL

However it is well known that light does not disperse through a parallel sided prism. So how can this interpretation or model of what happens be explained! Perhaps this model is wrong!

It is well known the dispersion does not occur to light traversing parallel sided prisms. Therefore if refraction at the first interface really is a function of wavelength then somehow it must recombine by the time it reaches the exit interface. This would require light to somehow differentially bend in arcs through the prism so that all colours (photons) can meet at the same exit point, regardless of wavelength. This can't happen as light can't bend differently within the media as this contradicts the straight line theory.

Does light refraction behave differently because of the type of prism involved and hence due to the different angles of the two interfaces? If this really happens there must be some way of informing light at the point of entry what type of prism it will be propagating through or what the exit angle will be. This is clearly ridiculous (not a very scientific term but you know what I mean). It would need to rely on some sort of "look ahead" capability or backward communication between the photon and itself at an earlier time or some other form of spooky action at a distance all of which in my view are completely unrealistic, quantum physics or not.

So is refraction at an entry interface really always just function of wavelength and entry angle and if not then how does dispersion occur only at the exit interface and then only if the angles are different? I believe the correct model for light traversing a rhombic (parallel sided) prism is shown in figure 6.3 below. This shows a light beam or single narrow light source (shown in grey on the left hand side) entering a rhombic prism at an appropriate angle just as in the triangular prism case. But this time dispersion or wavelength dependant refraction does not occur at the entry or the final interface and white light traverses the prism and emerges as white light as expected. I have conducted this experiment to confirm this result. A dotted line is included to show how the prism may be considered as two triangular prisms closely connected. This situation is completely reversible and light can go in either direction for the same effect.

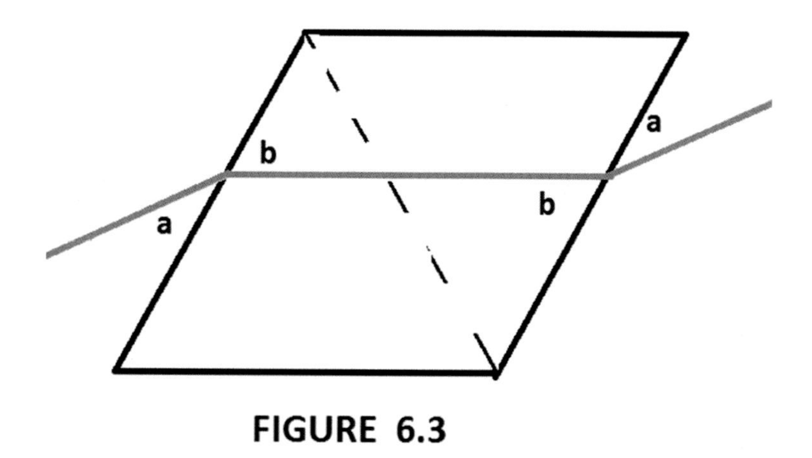

FIGURE 6.3

RHOMBIC PRISM
REFRACTION (NO DISPERSION)
CURRENT STANDARD MODEL

How can the current model of refraction and dispersion be explained when it shows light passing through a parallel prism behaves differently to that traversing a triangular one? Differential refraction and dispersion is not observed when light passes through a rhombic prism or any parallel sided transparent material. If a narrow beam of white light enters a parallel sided prism then white light leaves the other side after undergoing wavelength independent refraction or bending at either interface. Therefore the interpretation of what may happen in a rhombic prism as shown in figure 6.2 must be incorrect. And the interpretation of what happens in a triangular prims as shown in figure 6.1 must therefore also be incorrect!

I have conducted many experiments with prisms to investigate various situations and can confirm that dispersion never occurs at the entry interface in any situations I have investigated. It may occur at the exit interface if the entry and exit angles are not equal. But if the entry and exit angles are equal there is no dispersion! The most surprising case is when white light enters a triangular prism (equilateral or isosceles?) and undergoes internal reflection. It then emerges as a white beam without dispersion as shown in figure 6.4 below. I have conducted this experiment and simple geometry will confirm the equality of all angles shown, especially entry and exit angles "a". However it should be noted that the reflection process changes the photon orientation property so that the exit angle now needs to be the reverse of that in a rhombic prism to ensure dispersion does not occur.

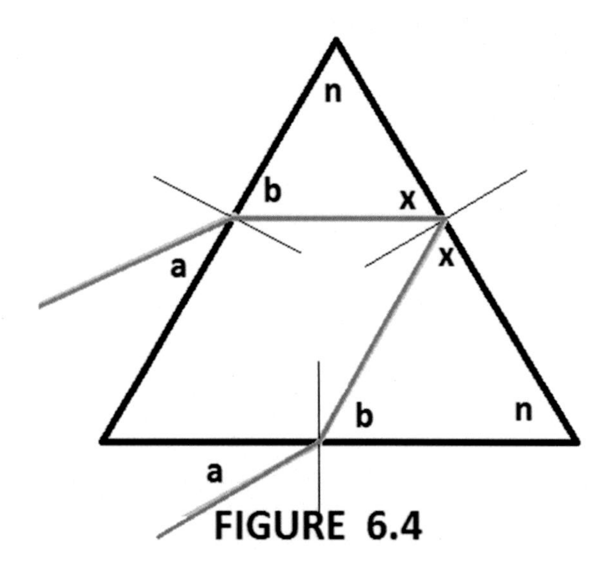

FIGURE 6.4

TRIANGULAR (ISOCELES) PRISM
REFRACTION AND REFLECTION
NO DISPERSION

This is what really happens. I have tested it! No differential wavelength dependant refraction or dispersion occurs at either interface if the entry and exit angles are equal. How surprising!

The above diagrams (figure 6.3 and 6.4) show how light travels through media, such as a rhombic prism or an isosceles triangular prism with internal reflection. For both situations all light leaves at the same angle it enters for all wavelengths. The prism only causes a deviation in the path the light takes. Any light travelling through the prism entering at a given angle in a narrow beam is refracted (bent) by the same amount at each interface. The refractive index is the same regardless of wavelength but at opposite angles at each interface. Because the refraction (bending) is the same for all wavelengths, no dispersion occurs. If white light enters then white light emerges. Both these situations are completely reversible. The same non dispersive action follows regardless of which direction the light is travelling. But is this always the case?

For white light (or any wavelength mixture) there may be transit time differences for each of the different types or "colours" of photon. This may be due to the speed differences through the media due to wavelength. This behaviour would not generally be noticeable as although the light (photons) that leaves at any instant may not be the same photons that enter, it would be the same relative mixture. Perhaps it is made up of different photons than those that entered due to wavelength dependant delay differences but this is not relevant for any refraction analysis. Any colour based differential delay in any transit media will have no visible dispersive impact on white or any colour light if there is no differential refraction. This situation of different photons leaving than those that entered would of course apply to all situations where white light is observed to traverse a prism.

I have used my own set of prisms and conducted many experiments to investigate these effects. In some cases I used a very narrow slit in opaque material covering one side (either on the exit and entry sides) of a triangular prism to test for dispersion. However I have not observed any dispersion or differential refraction based on wavelength at the first interface in any situation or at the last interface where the entry and exit angles are equal. Wavelength dependant refraction also does not occur at either interface in a rhombic (parallel sided) prism. Therefore there is no dispersion and no rainbow is created at either the entry or exit interfaces. Unfortunately I have not yet tested non isosceles triangular prisms where the exit and entry angles may not be equal but believe that dispersion may occur in such situations. Another area for more research.

So what is behind this behaviour? What is the correct model? How does wavelength dependant refraction and dispersion occur only at the exit interface and only if the angles are different?

Snell's law which defines the relationship between entry and exit angles of light traversing media interfaces supposedly applies in all cases. But under Snell's law refraction is not usually a function of wavelength. The above examples of experimental results prove this. So this law must be incorrect or at least incomplete as it doesn't describe dispersion correctly. There must be a missing component in refraction to explain these various situations and in particular to fully explain why exit interface dispersion only occurs if the entry and exit angles are different.

What must really happen with light traversing a triangular prism or in a general case where entry and exit angles may not be equal is shown in figure 6.5 below. In a special arrangement with this situation, the entry and exit angles may still be equal in magnitude but they are of the opposite orientation and there has been no internal refection to cancel this effect.

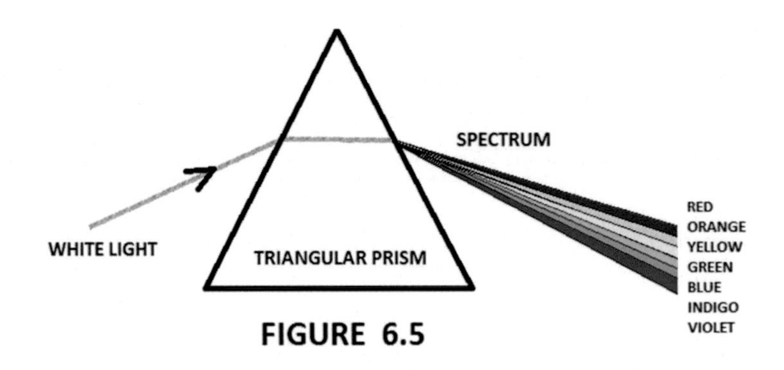

FIGURE 6.5

DISPERSION AT EXIT ONLY
NEW STANDARD MODEL

My interpretation of how dispersion occurs only at the exit interface and only in special situations! But why does light behave like this?

What becomes obvious for the situation shown in figure 6.5 and perhaps for all dispersion cases is that dispersion may not be reversible. If dispersed light (i.e. white light that has been broken down into a spectrum) was refocused to converge at a point on anther identical media interface but in an opposite way, what type of refraction would occur? Would all wavelengths undergo the same amount of refraction as happens in any initial media interface and hence not recombine? Or would they undergo reverse wavelength dependant differential refraction and recombine into white light after the first interface? But then what would happen at the exit interface? Would they undergo wavelength dependant refraction and dispersion again?

It seems that a rainbow of dispersed white light may not in general be readily combinable into white light through a prism. One likely outcome is shown in figure 6.6 below. This shows the original white beam together with separate red and purple beams (of very narrow wavelength) all entering a prism in parallel. Hence they all have the same angle of incidence at the entry interface. Again assuming the same effect as for white light, each beam of light must undergo the same amount of refraction at the first interface. Therefore the red light beam should bend by the same amount as the red light in the white beam and the purple light beam should bend the same as the purple light in the white beam. So, after the first interface all light must traverse the prism in parallel beams (shown with arrows). At the exit interface the same degree of wavelength dependant refraction should occur and the two monochromatic beams should bend differently and emerge in parallel with their white beam counterparts and hence diverge.

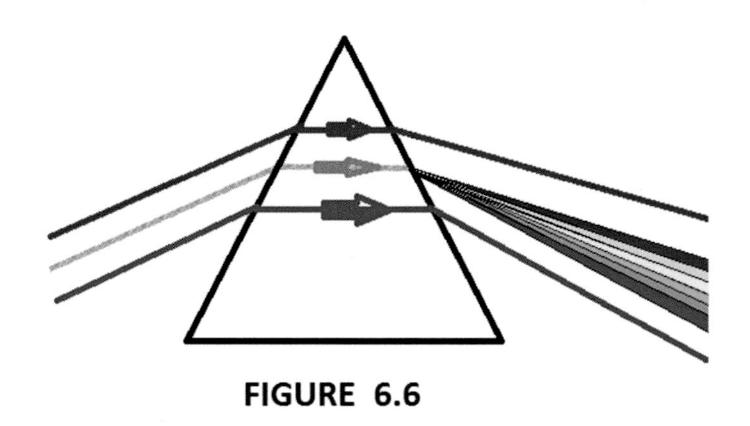

FIGURE 6.6

TRIANGULAR PRISM - 3 PARALLEL LIGHT BEAMS
MONOCHROMATIC RED (TOP), VIOLET (BOTTOM)
WHITE IN CENTRE WITH DISPERSION
SAME COLOUR ENTRY AND EXIT ANGLES

This general rule should apply to all light unless there is a very strange interaction between photons and the media because of the mix or spectrum of light. This would imply that light behaves differently depending on whether it is a monochromatic beam or in a mix with other colours. If so, then there is some explaining to do but I doubt it. I believe the light behaviour shown in figure 6.6 must be the correct model. But perhaps the quantum theorists or wave specialist already have another explanation and I look forward to finding out about it.

In the reverse situation as shown in figure 6.7, the red and purple monochromatic beams are traversing the prism in opposite directions to the white beam. They go in parallel with their counterparts in the rainbow but in the opposite direction. Therefore they are not parallel to each other but are converging. At the first interface (the old exit interface) both beams should again undergo refraction based on the angle of incidence but not wavelength as per the above discussion in figure 6.5. Each beam would therefore undergo a different amount of refraction according to Snell's law because they have different angles of incidence. But they could never be parallel within the media as this would require equal angles of incidence or differential refraction, both of which seem contrary to basic physics. Hence they may not recombine into white light at the second (old entry) interface if they are initially converging and therefore have different incident angles.

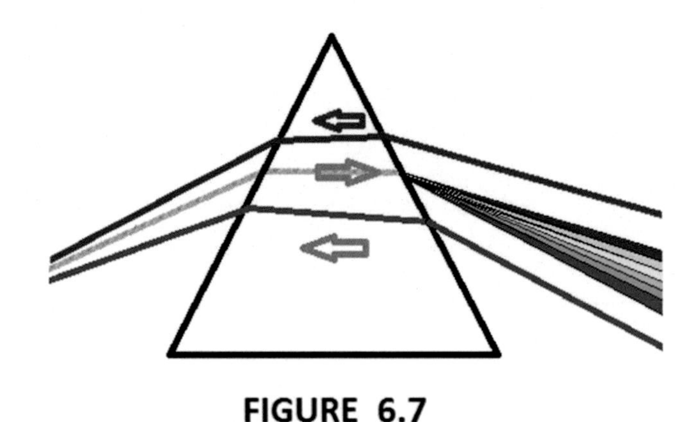

FIGURE 6.7

TRIANGULAR PRISM - 3 PARALLEL LIGHT BEAMS
RED (TOP), VIOLET (BOTTOM) - RIGHT TO LEFT
WHITE IN CENTRE WITH DISPERSION - LEFT TO RIGHT
DIFFERENT COLOUR ENTRY AND EXIT ANGLES?

So what really happens? Are the monochromatic beams parallel when they exit the prism in reverse of the discredited model of 6.1 above? Beams of monochromatic light should undergo the same amount of dispersion at both interfaces that the same colour in a white mixed beam would. They could exit in parallel but may not be co-linear. Hence white beam dispersion perhaps cannot be reversed. More study is required. This type of reverse dispersion experiment was supposedly carried out by Newton to confirm his theory that white light is a mixture or combination of colours. I am not so sure if he was really able to do this experiment or if he just thought that it was possible. To my knowledge, it hasn't been repeated anywhere else either. I have tried to do this experiment but unfortunately lack the necessary equipment to produce a converging dispersed (rainbow) beam. But from the above discussion it would seem problematic.

A PHOTON VIEW

To try and describe what may be happening I will use photons and simple mechanical concepts. I have not carried out any mathematical analysis of these situations using wave mathematics based on wave groups or other wave based theories. Also this mechanical photon approach is not based on complex quantum theory for bosons which applies to photons or any general quantum theory. But hopefully it explains the situation in general terms and should help improve understanding of what may be happening. These new models and any new theories based on them will need to be re-examined in due course and more detailed theoretical explanations of what may be happening will need to be produced. But in the meantime, here goes.

What really happens from a photon point of view? All photons of any wavelength with the same angle of incidence must undergo the same amount (angle) of refraction or bending at the first interface and follow the same path through the prism to the exit interface. Here may be a speed difference but this property is addressed later. And all photons of a single wavelength entering any prism at the same angle of incidence must also travel the same path through any prism and then exit at the same angle. In this case they all travel through the prism at the same speed. Wavelength dependant differential refraction and hence dispersion of a photon does not occur at the entry interface. It can only occur at the exit interface and then if the entry and exit angles are different. But how can each photon bend by the same amount at both interfaces regardless of wavelength if the entry and exit angles are equal or bend differently according to wavelength if the entry and exit angles are not equal?

Photons travel at different speeds through a media depending on wavelength, with red light travelling slower than blue light for example. But speed difference which may cause refraction differences and hence dispersion at the exit interface, is not sufficient. In particular it doesn't cater for entry/exit angle differences. So there must be another property imparted on each photon at the first interface to capture the entry angle. It must be related to the refractive index of the media and also be a function of wavelength (photon size). But to allow for all possible crystal alignments it can't just be a factor of the angle of incidence relative to some type of internal angle of atomic crystal alignment. All possibilities of photon entry and refraction angle as well as all manner of media atomic (crystal) alignment need to be catered for.

This property or change in photon behaviour would control photon behaviour through the media and the amount of refraction or bending at each interface. All photons would be refracted by the same amount at the first interface according to Snell's law using the refractive index of the media. There would be no dispersion at this point as all photons would be travelling at the same speed and not yet have any special media dependant property. If the entry and exit angles are equal but opposite (figures 6.3 and 6.4), all photons, regardless of speed or wavelength would then undergo the same amount of refraction at the exit interface. Snell's simple law would apply in reverse. But if the angles are different (figure 6.5) the entry angle and wavelength dependant factor imparted on each photon would result in wavelength dependant differential refraction at the exit interface creating dispersion and a rainbow. Snell's law would be incorrect or at the very least require a factor to cater for wavelength as well as angle difference.

Each photon (flat disc) would receive some distortion or "wobble" or other type of transmission behavioural deviation at the first interface. It could involve a change in either of the two field vectors, E or B with one or the other (most likely the E field) being somehow "modulated" by the media entry "event". Perhaps the change to each photon is caused by how it interacts with the atomic structure of the media according to size. Interestingly this may be related to an electrical property of the media such as the dielectric constant as well as being related to the refractive index. A possible relationship between these two properties also needs to be investigated. Perhaps some atomic property of a media that has yet to be identified may also be relevant. This new photon factor would then be used in the refraction process at the exit interface together with the exit angle. If the angles are equal (opposite?) then the factor would not apply to refraction and no dispersion would occur. However if they are different then wavelength (size) dependant refraction and hence dispersion would occur. The factor would need to be removed or cancelled out at exit so it does not apply outside the media.

The most likely explanation is based on simple physical size aspects using a simple almost mechanical model. The refractive index of the material must be directly related to the crystal structure or the atomic array "pattern" of the media. There may be some interaction between the atomic (crystal) structure of the media and each photon based on relative size and angle of incidence, similar to how X-ray crystallography works. This interaction would cause identical initial bending or refraction. There must also be some type of change in the dynamics or "geometry" of each photon at the entry interface that captures angle information, if that is possible. This property need extensive further investigation.

The crystal structure of glass or any similar transparent media has regular close atomic spacing. The estimated distance between atoms in glass is of order 10E-9 to 10E-10 metres. From my previous discussion on what a photon may "look" like I have suggested it may be like a spinning flat disc with no longitudinal dimensions. The "size" or lateral dimensions or diameter of a photon disc is estimated as being the same order of magnitude as its wavelength. Visible light has a wavelength of 10E3 to 10E4 Ang. So a photon would therefore be of order 10E-6 to 10E-7 metres dia. Red photons would be larger than violet ones by about a factor of 10.

Comparing these two dimensions shows a significant difference in the relative "size" with photons being approximately two orders of magnitude larger than the atomic spacing of the media. Therefore each photon "sees" of the order of thousands of atoms when it enters a media like glass. It also seems likely that larger photons "see" more atoms than smaller ones. That is red photons see more atoms than blue ones. Therefore perhaps red ones have to "work harder" to penetrate the media than blue ones. Perhaps this is a simple explanation for transit speed difference of photons through a media as an inverse function of wavelength.

One current interpretation of light being slowed down as it traversing a media like glass, assumes the media captures each photon and then re-emits it at a later time. But it would seem unlikely that each photon would favour any one particular atom in this initial encounter as a photon would see many atoms at once. Therefore it would be unlikely that capture and re-release occurs for any photons at all. Capture and release behaviour would also lead to transient energy capture by the media from the light which would cause an increase in temperature which doesn't seem to happen. Light photons are therefore unlikely to be captured, just transmitted but "affected" in some as yet unknown way.

But what causes dispersion? At the entry interface all photons undergo the same refraction or bending. But they also must obtain some form of distortion or "wobble" based on the angle of entry. Smaller photons (shorter wavelength such as violet) may get more "wobble" than larger ones (such as red) or perhaps the size (wavelength) causes larger ones to obtain a different type of "wobble" than smaller ones. If the exit angle is equal to the entry angle this differential photon "wobble" somehow ensures all photons undergo the same amount of refraction. But if there is a difference between entry and exit angles, the wavelength dependant "wobble" causes differential refraction and hence dispersion. Perhaps a combination of speed and photon "size" as well as difference between entry and exit angles can be translated into a new factor to include within the formula for the refractive index of a media. The wobble factor would be added at entry and removed at the exit interface. Perhaps a Parkes law modification to Snell's law is required.

An experiment based on using monochromatic light sources (multiple different types of lasers) could be set up to examine these situations in more detail. The amount of interface refraction caused by entry and exit angle differences as well as wavelength factors could be identified with appropriate experiments. Such experiments could also possibly identify likely candidates for other media dependant properties involved in refraction and dispersion and in creating any distortion or "wobble" that may apply. At this stage I have been unable to propose any specific mechanisms or parameter as well as values for this new property so Snell's law modifications will have to wait. I am also uncertain what types of media properties may be relevant to determining this new factor.

Perhaps the ε & μ factors used in deriving the speed of light will be involved. To confirm my ideas will require extensive research and theoretical analysis but at present I don't have the resources, ability or time. Watch this space.

DIFFRACTION AND INTERFERENCE

Some of the most fascinating and illuminating experiments about light involve the creation of interference patterns. They usually involve a strong single light source and various systems of lens, mirrors and panels with holes or slits in them. Yes it's all done with smoke and mirrors. Alternating dark and light patterns of light intensity are produced by particular combinations of the various devices. These types of experiments were used in early science entertainment shows. The outcomes are readily interpreted using simple wave behaviour as seen in everyday water wave situations and explained by wave theory.

Many experiments have been conducted into the phenomenon of light interference. Names such as Young with his double slit interference experiment and Fresnel with his lens interference experiments come to mind. A growing body of mathematical wave theory for light behaviour was developed by great names such as Huygens. Perhaps the most famous of all is the dual slit experiment of Thomas Young. This experiment is easy to set up and is very popular in physics books and demonstrations for all ages.

Young's experiment uses a source of light and a series of sequential panels. The first slit creates a narrow beam of light that spreads out due to diffraction which is the property of light to bend around a corner. It then also causes some other "special effects" such as "correlating" the light. This slightly diverging beam is then directed at the second panel with two very close slits, causing two similar narrow beams which also spread out. Again these two slits cause some property to be imparted on the light photons traversing either such as some angled "alignment" or further correlation but in an opposite way. The third panel captures the light photons that results from the interference of these two beams. These photons arrive in varying intensities at different locations and produce an intensity pattern on the detector panel.

The basic layout of the experiment is shown in figure 6.8 below. Light travels through the holes and produces a pattern on the detection screen showing variations in light intensity. The intensity of light on the screen varies from low (no light) to high intensity light on a periodic basis. The periodicity is a function of distance between second and third panels and second panel slit separation. There is no wavelength dependant dispersion. This behaviour is similar to that observed when water waves pass through a similar set of restrictions. This experiment led to acceptance of wave theory as the complete explanation of all light behaviour.

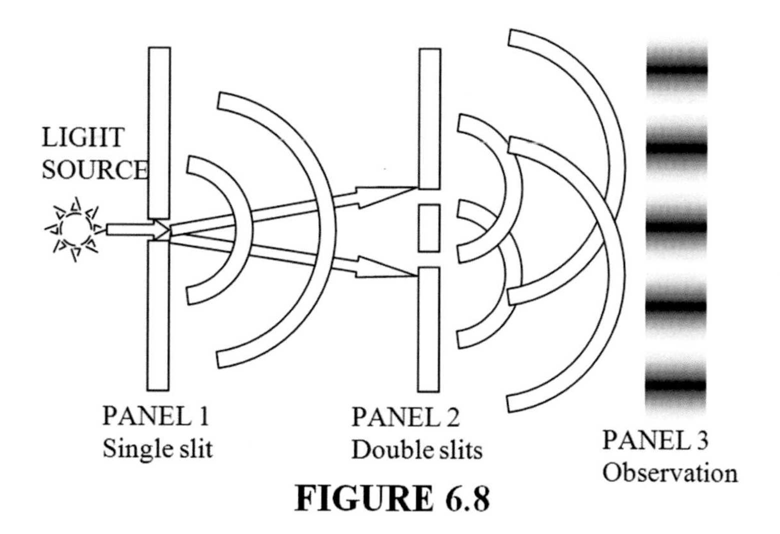

FIGURE 6.8

YOUNGS DOUBLE SLIT INTERFERENCE EXPERIMENT

Young's experiment has also been carried out using many different types of light and even particles such as electrons, which, according to quantum theory, supposedly also have wave like properties. They always produce the same interference results. The experiment has also supposedly been carried out using "single photons" which also exhibit similar interference results but I have serious doubts about this "result".

The current explanation is based on wave mechanics. It states that slots from panel 2 produce two sources of "correlated waves" of light that spread out from each slot. When these two waves arrive at the detector panel from the two sources they interfere. When they are in phase (whatever this means for light photons), either two peaks or two troughs, they add together and produce a stronger wave peak (brighter light) or deeper trough (also brighter light?). But when the wave peaks from the two sources are out of phase they cancel each other out and produce no wave or no light and hence darkness. All very simple if a light wave based approach is adopted. But what happens when a photon based approach is applied?

Now using a photon based interpretation of this experiment, bright regions are simple to explain. They simply involve more photons arriving at the bright points than at other points on the detector screen. It is simply the addition of energy of all the photons that are arriving over a period of time (higher photon rate). But what I find most challenging in understanding the results of this experiment is the explanation for dark regions. In wave theory, phase is a simple concept and if the waves are out of phase they supposedly cancel each other out causing no waves or dark regions. But if quantised light theory is used (i.e. photons), such an explanation for this experiment is problematic at least for me.

Dark areas are supposedly caused by light particles or photons arriving "out of phase" therefore cancelling each other out. Without really addressing the question of what phase means for a photon, there is a fundamental problem with this interpretation, what happens to the energy? Each photon has energy and even if a photon has some property called "phase" the energy still exists. It is not simply negative energy. So when any two photons arrive together (simultaneously?) both have positive energy. And even if they are out of phase, whatever that means in energy terms, they can't simply cancel each other out. This situation also creates another challenge for photons arriving together and that is the problem of simultaneity.

When photons from the two sources are supposedly completely out of phase how can they completely cancel each other out producing no energy at the meeting point. The fact that the energy of the two photons somehow disappears seems very problematic! Photons even with some form of phase difference are not like particles and antiparticles which supposedly can cancel each other out because one has positive energy and the other has a property called negative energy (or the quantum speak term for this property) . Where does the energy go? If different photons from the two sources arrive at a point on the detector panel at the same time (in phase or out of phase?) surely the energies would simply add together and produce brighter light.

If each of the two sources in panel 2 send out photons equally in all directions and phase interference is used to cancel some of them out there would also be a net deficit in energy. This surely can't happen. How can photon energy cancellation occur? How can the real energy of two photons cancel out to no energy just because they have some difference in behaviour? This seems to contravene the conservation of energy law. And what does out of phase really mean for a pair of photons? Is it due to differences in the phase of the rotating fields of each photon or is it due to a timing difference or what? And if phase is a challenging concept then how are dark regions produced? Perhaps dark regions can only means no photons can arrive at all!

The only reasonable outcome is simply that all photons that get through either of the second panel slots must hit the receptor panel or screen in some place or another. All photon energies must be accounted for. There can be no photon cancellation and loss, regardless of phase. That implies that photons just don't arrive at a minimum or null or dark point on the observation panel, in any phase or with any other property. They simply can never arrive at those locations, but why not? What property makes them avoid such locations or steers them away?

This problem is supposedly addressed by a new quantum model of interference. It says that each individual photon goes through both holes at once! In fact every photon supposedly goes in every direction, like Feynman's summation of infinite paths theory. When versions of the same photon catch up with each other again they somehow add or subtract depending on the distance they have travelled or other factors. But I have considerable difficulty with this proposal. It contravenes energy reality. If each photon is split into two (or is it more) then each should have the same wavelength or colour and hence energy. So the energy is doubled or becomes infinite for the infinite path approach! This is similar to spooky "action at a distance" which to me is Emperor's new clothes material again. It seems to add unnecessary complexity and reminds me of the epicycles approach used to defend the old earth centred cosmic model. The mathematics seems impressive and it may get the results but is it meaningful or correct? Surely there is a better more realistic and even simpler explanation!

HOW WOULD PHOTONS DO IT

What is the process by which photons interfere or interact with each other to create interference patterns? All photons coming through the first slit spread out to a certain extent due to diffraction. It can be assumed that there is no unusual pattern to this initial spreading out effect apart from perhaps a gradient in the intensity of the photons as a function of the angle of deviation from the normal straight line path. The distribution of photons would be something like a bell curve with maximum intensity in the middle and a gradual and equal reduction either side of the centre. If positioned equally either side of the first slot beam, each of the second slits would then receive a similar amount of light intensity or photon count. But the photons at each of the second slots would have an opposite angular difference or bias.

Some property must be added to each photon due to diffraction as a result of traversing the first slit. This may be similar to the type of angular "wobble" imparted on a photon in the refraction case above. There would be an equal amount of light but with opposite angular direction arriving at each of the second slits compared to the direction of light at the first slit. This would then cause an equal but opposite secondary distortion property for the light being diffracted by each of the two second slits. The orientation or distortion of the photons as a result of diffraction by each of the second slits would be similar in type of magnitude but opposite in orientation or angle. Each of these two slits would then create a similar diffraction and spread of photons with similar (bell shaped) intensities but with opposite angular properties or "wobbles".

Photons from both these second slits would then spread out due to diffraction and move toward the screen. If one slot was covered, the other would create a (bell shaped) distribution of light intensity on the detector screen. The angular (wobble) property would not be readily noticeable but may be detectable with special technology. The dark/bright intensity pattern can't already be in the distribution of photons coming out of either of the two second slits, quantum theory or not! Therefore it must be created by some form of interference between photons from either slit when both are open. The resulting interference pattern is the well-known wavelike variation in intensity shown in figure 6.8. At some locations the light intensity goes to zero while at others it is double the average.

According to current light theory (which I support) photons are made up of oscillating and mutually inducing E/M fields and have quantised energy and momentum. So in a sense if they have momentum they also have mass. What is the effect of this mass on the interference process and what are the ensuing forces if any? And what is the effect of these fields in the interference process? It is known that fields can "feel" attraction or repulsion. And "particles" with mass may collide. Therefore it seems likely that photons may have a mass collision or a field based impact on each other in special situations. Photons may "feel" attraction or repulsion with other photons if they are suitable aligned and in a suitable location or position when they "cross paths". This behaviour does not need mystical entanglement and other "action at a distance" theories that seem to be a basic part of any quantum solution to this behaviour.

So what happens to the photons that are headed for a dark region but don't get there? My theory is that they interact or to use the correct term interfere with each other because of an appropriate "meeting". This timely meeting then causes a collision or deflection that changes the direction of either photon to ensure that no photons reach certain locations (dark regions). Instead they are directed toward other (bright) regions. What seems to be happening is that each photon headed for a dark spot "bounces off" a suitably timed and aligned photon from the other slit that is also headed to a dark spot, perhaps not the same dark spot. It seems that pairs of photons are always set up to undergo deflection depending on the path travelled through the slits. This "matching" of normally unconnected photons would result from the property (wobble) imparted on each photon by the path taken. This experiment also only produces intensity variations, not dispersion and rainbows so it can't be a function of wavelength and must be the same for all colours.

For photons heading for a bright spot from either slit, perhaps there is no interaction with any other photons and they continue on without interference. But what causes grey areas? The degree of interference would obviously be a function of path length, just like in the wave based approach. Therefore some minor differences in the timing of deflection would produce differences in the amount of deflection. It could also be caused by the closeness of photons in each photon pair interaction. If they are very close they deflect each other more but if not then less deflection would result. The amount of photon distortion or "wobble" caused by traversing the first two slits would be part of this probability of interaction. The situation seems very mechanical to me and perhaps that is a satisfactory explanation at this stage. In due course formal mathematical solutions involving light intensity (photon count) probability density distributions, path lengths, delay difference distributions and the "wobble" factor may provide the complete answer. Then perhaps my photon interaction idea will be verified.

IN SUMMARY

This chapter shows that photon interactions can perhaps provide insight into all types of light behaviour. This could be used to update some of the current theories and models of light behaviour based purely on wave theory. Light behaviour such as refraction, dispersion, diffraction and interference which is currently explained by wave theory can also be explained by the particle theory of light using photons as flat discs of energy. These new interpretations may also lead on to new concepts and theories for light and identify other ways light may behave. In some situations light behaviour may have been incorrectly interpreted or explained by wave theory. Perhaps the photon explanation is better than the wave explanation in all cases and may avoid such misinterpretations.

As part of my particle theory approach I have proposed that a photon may have another property such as some distortion or "wobble". This may be imparted on each photon as it transverse a media interface as in refraction or as it passes a boundary as in diffraction. It may be related to or due to media differences, angles of entry and exit, and other transmission effects. Such distortion may be the key factor in controlling further behaviour. However, I have been unable to determine any specific type of photon behaviour or any media dependant parameters that may be causing this photon based behaviour. And I have been unable to fully derive any mathematical methods and hence prove my photon interaction proposal for refraction, dispersion or diffraction based interference. But I have clearly exposed some weaknesses in current theory and proposed the basis for new models of these types of behaviour. This alone should be sufficient to trigger interest and perhaps start a search for better explanations for these phenomena.

CHAPTER 7 –IMPACTS, OUTCOMES AND BENEFITS

So what happens now? What will be the outcome of my efforts if any? How do I progress my ideas and what impact will they have on physics? Well the first step was to write this book and I have done that. The next step is to try and get people, especially the right people with influence in this area, to read it. That is perhaps the most difficult and costly task. Hopefully I can get enough coverage so that it will be read by some armchair experts in the scientific community and even some real ones. Then with any luck someone may think some of my new ideas may have merit and be worthy of further consideration. But if I can't get my ideas and new theories circulated and acknowledged through this book then I will have to raise them in another way. The internet could be a platform but the problem is that it is mainly for popular junk culture and quick fix entertainment or is it? Perhaps there are scientific forums on the internet for my ideas.

Will my ideas and new theories suddenly change things in any physical sense? Well the answer to this question is obviously no. Life as we know it will not change. My new theories won't stop the earth orbiting the sun or the sun from shining or anything else for that matter. As for possible impacts on physics theories and models, the answer will depend primarily on the correctness, verification and acceptance of my ideas. But I am sure that when some of my proposals are scientifically tested they will eventually be proven to be correct. Then they will have a significant impact on the physics of light and gravity and on science in general. I believe some of them will change the way physicists view the two extremes of physics, the atom and the cosmos. Hopefully at the very least they will also impact some of the standard models of physics.

They should also lead on to the development of new research fields for light physics and also the development and application of new light technologies that will improve our lives? I believe we are on the crest of the next (light) wave of amazing light science and technology. At this stage this is at most wishful thinking but In the meantime here are some of the areas of physics and related technology that I believe may be impacted by some of my ideas.

PHYSICS (CHEMISTRY?) OF THE (NEW) ATOM

My new theory of the atom with active protons may have profound implications for many areas of physics and chemistry. It has the potential to open up a whole new field of research into atomic physics. The initial focus of atomic physics (or chemistry?) was the electron. This created electron physics and chemistry which is the basis of much of our current chemical knowledge. When physicists started to seek answers to radioactivity and properties of isotopes this helped provide a better understanding of the atomic nucleus and especially the proton and neutron. This started nuclear physics and led on to the discovery of fission and eventually fusion which has only made very big bombs. Current atomic physics is shifting to the proton using high energy colliders. This charged particle is being accelerated to very high energies and used in the search for information on new particles and new quantum properties.

So how does the dynamic proton and PP light fit in? How will my new atomic model and my new heavy PP light impact the search for more atomic detail and new particles? And what impact will these new theories have on some existing areas of research such as the Higgs boson and quarks? I have created a new term called Proton Chemistry for this new field of physics. It looks like a promising area for research that should improve understanding of the atom. But before it is addressed here is a brief overview of some aspects of current chemistry.

ELECTRON CHEMISTRY

In order to address any new field of atomic physics or chemistry it is useful to first review existing atomic chemistry. The whole concept of chemistry is based on the elements found on earth and interactions that can be observed and tested. Elements are found in molecular form and usually involve multiple atoms. But some elements can be easily ionised to various degrees depending on how many electrons remain. Elements are usually in one of three forms, solid, liquid or gas. Transitions between these forms or phases are well understood physical processes, based on the particle theory of matter and thermal mechanics.

An extensive range of atomic material or elements has been discovered over time and are now defined in the periodic table. New elements have been discovered using the latest technology and some man made elements have also recently been added but these are usually very unstable. Elements may also interact in chemical reactions to produce a range of outcomes and form various compounds. Some elements can combine to produce very complex compounds. Most chemical reactions are triggered by relatively low energies and low temperatures. Background radiation, mostly of natural origin and covering a range of wavelengths, is a factor in some chemical reactions. The sun is a major source of such radiation but the earth's atmosphere does a great job in filtering out some of the shorter more energetic wavelengths.

In any element, the proton count (atomic number) which is the same as the electron count, defines the position of elements in the periodic table. The electron count is also related to chemical behaviour or activity and determines the classification of elements into various types such as metal, semiconductor or insulator. Atoms with a "surplus" of electrons in the final "shell" are usually metals and are good thermal and electrical conductors. Those with a "deficit" are usually non-metals and are good insulators. Atoms with "completed" electron shells are called inert elements and they are very stable. The electron has a number of quantum states and rules which control how it behaves in particular situations. They include electron "orbital" positions which determine where each electron is in energy terms and how it can transition between energy levels or orbits. The exclusion principle also restricts electron behaviour.

The active component of common chemistry is of course the electron. The various types of electronic bonds within and between atoms and a wide range of electron energy interactions control these processes. The creation and capture of electron light is also involved in many of these chemical processes. It could be summarised as (relatively) low energy atomic and molecular interaction at low electron energy levels. It is well explained using the current (planetary) model of the atom. This electron chemistry is the basis for almost all the earth bound chemistry that we know and use in our technology today. There is sound quantum theory behind electron based chemistry.

Also associated with this electron "shell" information and electron energy level activity is an extensive theory of spectral analysis. There is a direct relationship between electron activity in terms of energy level (orbital) transitions and the related photon energy or wavelength. Electron energy increase (upward jump) is caused by energy capture from photons of the correct wavelength (creating absorption spectral lines). Electron energy release (downward jumps) creates photons at the appropriate wavelength or energy level (creating emission spectral lines). Either of these types of spectral lines are often used to identify different elements.

Early last century the atom was reasonably well understood and considered reasonable stable. Three nuclear particles were identified, the electron, the proton and the neutron with the proton and neutron believed to be contained in a static central nucleus. Only the electron was considered to be active or dynamic. But there were still many unknowns. Physicist wanted to explain radioactivity, isotopes and other mysteries of the atom that were believed to be due to the nucleus. So the nucleus then came under attack. The electron was considered to be indivisible but the nucleus with its protons and neutrons became the new target. Eventually a better model of these particles and how they combined into atoms was constructed, quantum of course.

By the middle of last century the "solid nucleus" was well and truly smashed._New high energy atom smashing machines produced some strange "events" which indicated the existence of other as yet unknown particles. Many new particles were discovered and a particle zoo was created. It was assumed most came from the two larger subatomic particles, the proton and neutron. The electron was and still is considered indestructible and there were no other obvious sources. Most of these new particles were not vary stable, not easily detectable and had very short lifetimes. From these results and associated new mathematical modelling based on quantum physics, it was proposed that the proton and the neutron may not be "solid" or fundamental after all.

Of the original three sub-atomic particles, the electron is still considered to be indivisible and stable (apart from some "special" quantum effects). However, some physicists proposed that protons are capable of spontaneous decomposition into other particles. Subsequent analysis has challenged this theory. Current atomic theory states that protons and neutrons are stable (mostly?) but are made up of even smaller parts. The standard model proposes that each is made up of three subcomponents called quarks but the types of quarks that make up each one are different. There are supposedly six quarks (three pairs) each with different quantum properties. And while quark theory is "accepted" theory for protons, re-examining the new atomic model with proton dynamics may shed new light on this. Perhaps the proton is not subdivisible and quark theory is just another way of explaining unusual observations (epicycles?).

NUCLEAR (NEUTRON) CHEMISTRY

Early chemistry was based on the stability of atoms, which it was believed could not be created or destroyed. But it was soon found that during experiments some atoms decayed and spontaneously transitioned into other elements with different atomic numbers. Little was known about such "neutron chemistry" until the neutron was discovered last century. Then some classes of elements were detected with similar chemical properties but with different atomic weights. The identical chemical behaviour was assumed to be because they had the same electron and proton counts (atomic number). The mass difference (atomic mass) was believed to be due to an increase or decrease in neutron count. The different types of a particular element were called isotopes. The identification of isotopes of elements with the same properties but different weights (i.e. with more or less neutrons) led to a new field of chemistry.

Some of these new elements, especially rare metals such as Radium (found by the Curies in pitchblende) showed new types of atomic behaviour such as glowing in the dark and (deadly) radioactivity. They also decayed into different lighter elements. Neutron chemistry began with the study of this radioactivity but soon led on to the study of nuclear decay and other nuclear reactions. The decay process, called fission, involves the breakdown of (heavy) atoms into lighter ones. It was the first nuclear chemical process to be discovered and opened up a whole new path in the discovery and creation of new elements. In certain circumstances some atoms supposedly capture neutrons. This causes atomic instability and breakdown releasing more neutrons. The reaction also results in the loss of relatively small amounts of mass and the release of relatively large amounts of energy per unit of mass, according to Einstein's mass/energy equivalence and the formula $e=mc^2$.

The fission process can supposedly occur to most heavy elements in the periodic table down to iron. But it is more likely with unstable isotopes of heavy metals. It was discovered that a particular (lighter) isotope of Uranium (U 235) could spontaneously breakdown into smaller elements with the release of more neutrons than are required to trigger the reaction. This process could potentially create a chain reaction releasing large amounts of energy in the form of radiation. This process was first created in a controlled way by Enrico Fermi in a lab in Chicago using some form of moderation to restrict the neutron chain reaction. The process was then used soon afterwards in an uncontrolled way to create the first atomic bomb. It is also used in a controlled way to create heat for power generation. But it also created a new type of manmade very deadly and hard to manage (long lifetime) nuclear pollution. Unfortunately while the atomic age began with a bounce, the nuclear age began with a very big and dirty BANG.

The next activity to be related to neutron chemistry was the fusion process. Fusion is the energy process of our sun and other suns. It was largely discovered as part of solar nuclear synthesis. The chemistry of the sun was analysed using spectroscopy to try and determine its energy creation process. Helium was found and it was assumed that it was being produced by a high energy Hydrogen reactions. Smaller atoms such as Hydrogen (or at least isotopes of Hydrogen with neutrons) were believed to fuse together to make larger ones such as Helium. This process is different to fission and involves atoms combining (fusing) together to produce heavier ones, hence the name nuclear fusion. Fusion also involves neutron capture and release in a chain reaction. Considerable energy is required to trigger fusion (fission is often used) but much larger amounts of energy are released per unit of reactant than in a fission reaction.

Both of these nuclear processes are reasonably well understood from a theoretical and a practical (mechanical) point of view. Fission of some elements occurs naturally on earth but at an extremely low level. I am not an expert on these processes but it is interesting to note that one of the keys to initiating a nuclear reaction is to use "slow" neutrons. Nuclear reactions produce high energy (fast) neutrons but can supposedly only capture slow neutrons. The other main requirement is a critical mass. A certain (minimum) amount of nuclear material in a specific configuration (shape) is required to initiate and sustain a chain reaction. There are many more requirements regarding the purity of the material, the triggering process and how to slow the neutrons down, but these details are beyond me and this book.

The nuclear industry is now well and truly established. Man-made nuclear (neutron) chemistry has resulted in the release of significant amounts of energy and pollution. Uncontrolled fission has already been used in bombs during World War 2 creating massive destruction. Fusion (uncontrolled), which is supposedly less polluting than fission (fewer radioactive by-products?) has been used to create Hydrogen bombs. Both have been developed, tested and stored in large numbers by "peaceful powers". These weapons of mass destruction are enough to destroy us all.

At this stage only fission has been "tamed" in any scientific way. Controlled fission (or at least controlled in most cases) is used in nuclear reactors for research, creation of radioactive elements for medicine and heat generation for power stations. It has also been used for powering satellites and driving naval vessels, removing the need for regular refuelling. However, if sufficient allowance is made for safety and pollution mitigation, it is very expensive to generate electricity from fission. In most cases it is more expensive than current non-renewable (fossil fuel based) energy systems although the long term pollution problems from these have also yet to be fully costed! Fission accidents (meltdowns) have already happened with disastrous consequences, producing deadly pollution and incurring untold long term costs.

Many attempts have been made to "tame" fusion but so far without success. The race (crawl) to produce "non-polluting, free" energy from fusion has incurred very large expenditure and resources but no real productive output. The search to copy the suns process but control fusion continues but difficulties with initiation, duration and stability indicate there is still a long way to go. Costly installations have not yet produced free energy, just confirmed that nothing comes cheap. Will the discovery of PP and other new light ideas be relevant? Will the creation of PP technology such as PP lasers help with fusion research?

PROTON (NUCLEAR) CHEMISTRY

As a result of my work I believe that a new field of physics or chemistry, called proton chemistry will arise. In contrast with electron chemistry and perhaps neutron chemistry, proton chemistry requires much greater energy. It is a completely unknown and new field but when it is developed it will open up vast new areas for research. But what is proton chemistry? Perhaps proton physics would have been a better term but you will soon see why I have tied it to chemistry. Well, proton chemistry is similar in a way to electron chemistry but instead of being driven by electron activity or electron atomic bonding, it is driven by proton activity and possibly something called "proton (nuclear) bonding".

Now immediately I can see raised eyebrows and questioning looks from experts as they exclaim "there is no such thing as proton chemistry let alone proton nuclear bonding! What is this guy on about? Well perhaps there is something called proton bonding and it will soon be discovered in this new form of atomic activity (chemistry?) that I am proposing. This proposal is so bold and new that even I don't have many ideas about it?" I am really going out on a limb with this one but I strongly believe it is an interesting area for research and will produce many new scientific outcomes. It is time to bring the proton back to life.

This new chemistry would need to be at a much higher energy and hence temperature level than that necessary for electron chemistry. The base state of matter for this new chemistry would therefore most likely be high temperature plasma. The components or inputs would not be atoms but stripped ionised atoms or parts of atoms. It will be completely different from electron chemistry in that it will most likely involve the creation or breakdown of the nuclei of atoms or at least change proton numbers in nuclei. The normal earth environment which is suitable for electron chemistry would not be suitable for proton chemistry. It is unlikely that any situations of proton chemistry can be observed naturally. But there is a significant potential for further analysis and research in this area through the high energy accelerators that are key to current physics research. Perhaps special high energy, high temperature environments could be created to investigate such chemistry. It may also be detectable by or related to some existing cosmic research.

I believe proton chemistry can proceed in two ways, similar to nuclear (neutron) chemistry. In some cases larger atoms or at least larger nuclei could be created from smaller ones (in a similar way to fusion). In others it can lead to the breaking up of larger atoms or nuclei to make smaller ones (like fission). These two processes are perhaps related to the size of the atoms involved, in a similar way to neutron chemistry. Perhaps it is closely related to or is an integral part of what we now know as nuclear (neutron) chemistry. Perhaps it is really just the same thing.

So if you thought atomic physics or chemistry was a relatively quiet area for a stable, non-challenging career, except at the extremes of size, energy and complexity, think again. My new theories of light and gravity will surely change all that. It will surely put atomic physics or is it chemistry, back in the floodlights. This idea is not the main theme of my work. It just arose out of extending the idea of proton light and the similarities between electron activity and possible proton activity, albeit under different circumstances. Of course this is all bold conjecture but I believe it is definitely worthy of serious scientific analysis.

DYNAMIC PROTONS, PROTON LIGHT

The proposal that protons are active and not static as per the current atomic model fits in well with my theory that electrons and protons have similar properties. Therefore they share similar processes such as "orbiting" the nucleus or atomic CG and undergoing energy level jumps and creating light. I now firmly believe my hypothesis that protons can and do create light and in particular a special bandwidth called gravity light (CGBR) is correct. Hence my idea for gravity based on light pressure (TOLG) is soundly based. It just needs to be proven.

I have been unable to obtain any information on my idea of dynamic or orbiting protons from any current sources or reference material and will continue research into this area. Perhaps this idea has previously been proposed and there are extensive theories on it and what it implies for atomic activity and light. However, it seems that current quark theory and the QCD model imply it hasn't as they directly challenge any such proposal. Research on this subject is most likely leading edge and hence any documentation may be very difficult to access. Also any associated theoretical modelling would be way beyond my mathematical capabilities at this stage. But it seems to me that such an idea, right or wrong, would surely be headline news by now if it had been raised so it must not have been raised before.

PP AND GRAVITY (TOLG)

Light created by proton activity and is the main focus of this book and is described in some detail in chapter 4. I have called this PP light for proton photons. While a sound scientific description and justification for this idea has been given, I have yet to fully and independently prove that PP light exists. Some areas for research based on high energy accelerators have been suggested but these are still not well defined. Perhaps some work currently being undertaken into theories such as the Higgs boson or new work into the new atomic model I have proposed may prove fruitful in creating PP. This research could have spectacular outcomes and I will definitely be watching this area with interest.

Gravity is caused by PP light pressure and shadowing according to my new theory of gravity (TOLG) set out in my first book. In this work I developed a basis for all aspects of gravity including the reason for the inverse square law and for a new type of equivalence. The main component of my gravity theory is this new type of PP light. I proposed that certain wavelength of PP occur naturally in the universe as background radiation. This is proposed as being similar to the well-known CMBR but at much shorter wavelengths or higher frequencies and energies. Therefore it is capable of creating much higher pressure in collisions with matter. I have called this background gravity light Cosmic Gravity Background Radiation (CGBR). Once PP has been created and measured in a controlled environment, the search for gravity light (CGBR) can proceed and when discovered, light gravity (TOLG) will be proven.

RED/BLUE SHIFT

The section on red shift analyses various possible causes for the well observed phenomenon of cosmic light shift. In particular I attempt to debunk the myth that light shift is only caused by relative radial movement as per Doppler shift theory. Other possible causes of either red or blue shift are considered but what really causes these phenomena in all situations is still a mystery. In particular a complete explanation of blue shift and to an extent red shift remains elusive. If they are really caused by relative radial movement, this implies some form of absolute movement detection by sender and receiver? Does this then imply a universal frame of reference (a type of ether?) or what? And if movement is behind red or blue shift this seems incompatible with the concept for light of single uncorrelated non longitudinal, lateral flat disc photons.

Other potential causes of shift were examined but in general no sound overall cause was found to explain all situations. In particular it also seems likely that there is possibly a combination of causes which combine to produce the overall effect. The current BBCM model proposes that all long distant large scale red shift is caused by expansion of the universe. This supposedly overrides any other shift that may be present in very distant sources. As I don't support the BBCM I don't accept this explanation. It also seems problematic because if everything expands, so do rulers, so who knows. Perhaps only space between atoms expands, not within atoms and not any particles. But this also seems a challenging concept (epicycle). Cosmic sunset could cause long distance red shift but surely this would have already been thoroughly analysed and allowed for. But are there other simple explanations that involve gas, dust and distance?

Perhaps there is similarity with our radar technology. Current radar technology sends out pulses of high frequency signals that bounce of an object. The received signals provide information on the object they bounce off. The strength of the bounced signal provides size information and its direction or changing direction provides location well as direction of motion data. The time separation of the returned pulses provides information on radial speed difference. Any supposed change in frequency of the returned signal (ala Doppler) is not used, only the intensity, direction and timing of the returned pulses.

So is there something to learn from this for red shift? Is light or are photons made up of a string of coupled particles not just one as currently proposed? The energy and wavelength may then be related to the quantity and separation of such a string or sub-particles. It may then be possible to explain shift as a change in the timing separation of the particles in a burst. Any change in separation could be caused by movement of the body that created the light or the one that captured the light or a combination of the effects of both. This would produce the required outcomes but the price to be paid is a complete change in the concept of a light photon as a single discrete particle.

And how does this new burst photon concept fit in with the current standard atomic model and electron energy jumps for photon creation and capture. Does an electron ring or oscillate when it creates a photon burst and is this the process that creates a string of sub-photons. If so then what determines how long it rings for and how many sub-particles are in the string or burst? And how does this all fit with Plank's and Einstein's theories? Well needless to say I am way out of my depth so I hope quantum theorists can investigate this late addition to my light work and come up with something more realistic but perhaps that is a quantum contradiction.

SPECIAL RELATIVITY

If gravity is caused by the pressure of PP light or CGBR as the new gravity theory proposes, then gravity acts at light speed. It also shows that nothing can move faster than light speed if it is due to the force of gravity. No gravity pressure (force) can act faster than the light pressure that causes it. But what about movement caused by other external or even internal forces? Can these reach speeds higher than light speed? This would require infinite energy according to S.R. so it is also not possible. Perhaps TOLG explains a lot about Special Relativity!!

The other aspect relevant to relativity is that any shadow that a body creates changes as the body moves. So if another body experiences a CGBR mutual shadow from a moving body, it is the shadow cast by the bodies when they were in other locations. The shadow which creates CGBR pressure/gravity is from a previous time and position. As bodies move, the mutual shadow they create changes depending on the speed and the distance between the bodies and the speed of light. This speed differences causes latent effects such as drag which increase as the speed difference increases. But any two bodies moving apart at greater than light speed (possible?) will experience no gravity force.

What this means is that there seems to be a concept of absolute movement or non-movement with respect to CGBR. If bodies are not moving relative to CGBR then their mutual shadow is also not changing. However if they are moving relatively to (or is it absolutely to) CGBR then the shadow is also changing. Of course the movement is related to the relativity between the speed of movement and the speed of light which is very fast but this new concept may have very interesting consequences and may explain some relativistic aspects. But what does this mean for red shift at greater than light speeds?

GENERAL RELATIVITY

The whole theory of General Relativity (G.R.) may be significantly impacted by new PP light and the new gravity theory based on it. The overall mathematical approach of G.R. based on mass distorting space/time may be similar to the concept of mass creating mutual shadows. But the full mathematical theory of mutual shadowing needs to be derived to confirm this. However, the basic concept of G.R. is that each individual body of mass (or amount of energy?) creates space/time distortion by itself which may then affect the motion of other masses. So if a space/time distortion detector could be invented it could measure distortion created by an individual body. But in TOLG shadow gravity, it takes two (or more) bodies to create mutual shadows and hence create a force of gravity. Therefore a CGBR gravity light detector would not be able to detect an isolated body. This is a significant difference between the two theories.

Perhaps there are many other differences that need to be identified and reviewed. However, at this stage I have not conducted any detailed research into other aspects of the GR theory may be specifically affected by TOLG apart from my view on "gravity" not bending light which is discussed later. One aspect that is of major concern to me is the so called equivalence principle which was supposedly what triggered G.R. This principle seems incompatible with causal theories such as TOLG, Newton's gravity and other theories. Newton related force "F" and acceleration "a" for any body of mass "m" in his famous formula, F = m * a. However it seems that he assumed causality was involved, not a direct "equivalence" between force and acceleration as per G.R. Causality is discussed later in the chapter on mathematics.

Einstein's equivalence principle, I believe, means that it is not possible to tell the difference between a body which has inertial movement in a straight line (or is stationary?) and no force is making it change or when a force is accelerating a body. It suggests that accelerating in a gravity field is the same as inertial motion, both bodies move along non-energy changing geodesics. The simple example that led Einstein to this idea was of a person falling and accelerating due to gravity. The person will "feel" the same as if he was floating freely in space where no force is being applied and no acceleration is occurring. The other example compares a person in a stationary lift experiencing gravity to one without gravity but being accelerated by a rocket of 1G force. The occupant can't tell the difference. This was interpreted as meaning there is none.

The problem I have always had with this principle is that it seems to rely on human sense or a concept called "feeling". Now we all know that human senses are often deceptive. Not "feeling" something is not what I would call a good proof of a scientific theory. The Ptolemy Earth centred universe was largely based on not feeling any movement on earth and hence saying there was none. In the dark ages with primitive religious beliefs, alchemy and witchcraft, people may have been blindfolded and ask to tell the difference between events. If they couldn't they were drowned or burned at the stake. So much for using human feeling as a proof of truth.

If an external force is only applied to one part of the body, such as on the feet, it will be felt explicitly. Bones and muscles and other bodily components then carry the force up and throughout the body but this is not directly felt. Downward pressure on all parts of the body due to gravity may not be felt but are still there. Only the (upward) force on the point of contact is felt. If in free fall toward or in orbit around another body creating gravity then a force is still acting on the body. It may not be felt but does this really mean there is no force? I don't think so.

In a free fall situation with gravity, the force (a directional difference in PP pressure in TOLG) applies to all atoms (subatomic particles?) in the body. There is no localised pressure or force only on one part of the body such as the feet and therefore there is no transmission of force through the body as in a static earth bound case. This is why it is not felt. It is still a force but it is unnoticeable. Now to say because it can't be felt means there is no force is perhaps stretching the imagination too far. A CGBR intensity "detector" (if and when one is ever built), may be able to detect directional intensity differences and hence pressure differences and a gravity force. But a human may not be able to feel anything. So does equivalence just mean that something can't be felt? Like "Ptolemites" not being able to feel the earths movement. However, I believe there is a more fundamental problem with equivalence based on causality and energy dynamics.

The real issue is; what are the net forces on a body and what are its movements? If there is a net force in one direction, what are the directions of movement and acceleration of the body? This is all simple (Newtonian) physics. Is the movement always orthogonal to the direction of the force? Of course for the force to have a direction and for the movement of the body to have a direction, there must be some reference frame for these directions to have meaning. But this is relativity. Is the only reference frame the body itself or are other bodies involved and if so how? Perhaps we can use Mach's idea of the universe as a frame of reference. But again let's proceed based on a general understanding of these factors and conventional physics.

Some basic situations are compared with or without forces, accelerations and changes in movement. These all assume a body can't accelerate without a force being applied and if a net force is applied then a body will accelerate. However, this concept may already be contrary to G.R. equivalence. Perhaps the real issue is can any differences between situations be observed and measured, not just "felt". If no difference can be found then are they really equivalent. Proving situations are different involves using parameters that are observable or measureable which in some cases is not as simple as it sounds. Here are some examples which are also summarised in the table below.

The first situation is when a body experiences a net force of gravity from another body but that force supposedly can't be "felt". If the force has a component (vector) in the direction of motion this results in acceleration in the direction of motion. It also involves energy transfer from the force to the body which has energy implications. Due to the laws of energy, this situation must therefore either be a transient situation which must have some termination point or a non-transient balanced situation. There are at least three cases.

The first case is when the bodies eventually collide. This will happen in finite time and be observable or measureable. Another is when the bodies pass nearby. Each transfers energy between position (field energy?) and movement (kinetic energy). However, energy is always conserved for each body and overall. This type of interaction has been used to manage the trajectory of many earth launched satellites. They have been given a "kick" by the gravity of a particular planet as they pass by to make them fly nearer to and observe other solar system bodies. A third is when one body passes through another and oscillates back and forth forever. Each of these situations should be observable, detectable and measureable beyond feeling.

A body bouncing back and forth through a hole in the larger body due to gravity is a special case of simple harmonic motion. This simple harmonic oscillating motion is similar to circular motion. Assuming no friction, this type of oscillation can continue forever and hence is not transient. Total energy is conserved and just transfers between kinetic and potential energy. And just like the orbital situations described before, both bodies must be moving or bouncing, albeit by different amounts depending on the relative masses of each.

The next case of a force that can't be "felt" involves one which is always orthogonal to the direction of motion. When the force on a body is orthogonal to the direction of motion, and the acceleration of the body "balances" this force, this is circular motion like an orbiting body. There is acceleration in the direction of the force but not in the direction of motion. Therefore there is no work done by the force and there is no energy transfer. It is a form of simple harmonic motion in a similar way to the bouncing body described above. This situation is just like the earth orbiting he sun. Again ignoring friction, it can continue unchanged forever as it involves no energy loss or gain. There is angular motion and of course momentum and stored energy as a result, just as there is for linear motion. But is the energy of circular motion, absolute or relative? No force or motion is "felt" and there is no way any difference can be detected by waiting for a transient situation to finish. But it should be possible to detect orbital motion of one body about another body using the universe as a reference.

The next situation is simply floating in space with no net force being applied (e.g. no gravity or PP light pressure difference or rocket engine or whatever). Again no force or motion can be "felt". The body may have "inertial" movement and as such have momentum and kinetic energy relative to another (inertial) frame of reference. But because there is no force, there is no acceleration and hence there are no energy transfer implications. Such a static (or moving) inertial body can supposedly continue in a state of standing still (or moving?) indefinitely. No energy is ever gained or lost or transferred between types or forms of energy. There is no way of detecting any movement unless another external reference body is used, just like for an orbiting body, but the universe is always there.

The final case involves force acting on a body and either accelerating it or not. In these situations the force can be felt but motion, if any, can't be felt. The first example is a body (person?) standing on earth (in a lift or not) and experiencing a force of gravity of 1 G downward. This is balanced by an upward force of 1 G on the body (feet) from the earth or lift floor. There is no net force and the body does not accelerate or move. The second example is a body in free space away from any gravity force but being accelerated by a rocket creating a force of 1 G. This is the only force acting and it causes an acceleration of 1 G on the body changing its movement or velocity. These two situations are equivalent according to G.R. but are they different.

In the static case no energy transfer or loss or gain is involved so it can continue indefinitely. This situation should be detectable as it can be observed that one body is standing on another (bigger rock) body. In the dynamic case of a body being accelerated by a rocket, energy is consumed by the rocket and transferred to increased movement (velocity and kinetic energy) of the body. This can only involve finite energy transfer and therefore must be a finite or transient event. The body (lift) will eventually stop accelerating and the occupant will eventually stop "feeling gravity" and eventually be able to notice the difference.

One implication of G.R. equivalence, I believe, is that equivalent situations should be able to continue indefinitely. They don't involve any net energy transfer and can't be transient and therefore they always seem identical. But the real difference at least between "floating " and "orbiting" is that one involves acceleration and circular motion (angular motion in general) and hence rotation energy and the other does not. The question then becomes can such acceleration and circular motion and rotation energy be sensed, detected, measured. This is another relativity question that has similarities to "Mach's bucket" or is it "Newtons bucket". I am not sure where this all leads but these explanations seem to challenge G.R. equivalence. More research is required into equivalence and the impacts TOLG will have on it. There should be sufficient material for another book! But perhaps not, just another chapter!

There are other aspects of G.R. that may need to be reviewed when TOLG is found to be correct. The major ones relate to solutions of G.R. and theories which have been derived from them such as the expanding universe theories and so called Big Bang model. Even Einstein seems to have had major doubts about some of these cosmic theories. Perhaps his eventual acceptance of an expanding universe was his "biggest blunder" not the cosmological constant, Λ, he inserted to prevent it from happening. How would Einstein have responded to TOLG and the above discussions on equivalence if they were proposed when he was developing G.R.?

The various situations used to discuss and analyse equivalence are set out in the table below. The main issue is; are there differences between the situations and are they observable, detectable or measureable in a finite period of time? Differences can't be just "not feel-able" for the situations to be equivalent!

EQUIVALENCE SITUATION COMPARISON TABLE

SITUATION PARAMETER	FALLING NEAR EARTH	ORBITING NEAR EARTH	FLOATING IN SPACE	STATIONARY ON EARTH	ACCELERATE IN SPACE
FORCE	YES 1G GRAVITY NOT FELT	YES 1G GRAVITY NOT FELT	NO FORCE NOT FELT	YES 1G BALANCED FELT	YES 1G ROCKET FELT
ACCELER'N	YES NOT FELT	YES NOT FELT	NO NOT FELT	NO NOT FELT	YES NOT FELT
VELOCITY	YES CHANGING NOT FELT	YES, FIXED, CIRC'R NOT FELT	YES OR NO FIXED, LINEAR NOT FELT	NO EXCL. ROT'N NOT FELT	YES CHANGING NOT FELT
TIME	1. TANSIENT 2. PASSING 3. SHM/OSC.	NON TRANSIENT (CONTIN.)	NON TRANSIENT (CONTIN.)	NON TRANSIENT (CONTIN.)	TRANSIENT FINITE
ENERGY	1. FINITE 2. BAL. 3. OSC.	NO ENERGY CHANGE (ANG. ENERGY)	NO ENERGY CHANGE	NO ENERGY CHANGE	FINITE TRANSFER
CAUSAL ACTION	YES DYNAMIC	YES DYNAMIC	NO STATIC	NO? STATIC	YES DYNAMIC
DETECTABLE	YES, TIME, OBSERV'N	YES OBSERV'N	NO?	YES OBSERV'N	YES TIME

THE BENDING OF LIGHT BY GRAVITY

One of the major outcomes of General Relativity (G.R.) was the prediction that gravity bends light. This became one of the most important "verifications" of G.R. A famous experiment to detect and measure light bending was undertaken shortly after G.R. was published. Using cosmic photographs or maps it is possible to observe the positions of stars near the sun during an eclipse of the sun by the moon. These could then be compared with their normal positions obtained at other times. The results of this famous experiment showed a very small change in location of light from certain stars near the sun during the eclipse. It was assumed this apparent position change was due to the deflection of the stars light by the suns gravity. This was a significant deflection in cosmic terms and was used to "prove" that G.R. was correct.

Of course Newtonian theory also predicts that gravity will affect the trajectory of a particle with "conventional" mass moving near another body. Calculations based on Newtonian gravity produce a solar deflection of about 8 arc seconds assuming that photons have mass, not just something called rest mass or whatever the current quantum term is. Einstein also originally calculated an 8 arc second deflection of light near the sun using his G.R. theory. But a much larger deflection was detected in the experiments (almost twice or about 18 arc seconds).

Einstein eventually doubled his calculations saying that light is bent twice as much as Newtonian gravity would suggest. Einstein supposedly explained doubling due to the combined effects of G.R. perhaps firstly warping space and then time. Perhaps this is because of the high speed involved. Light is of course travelling at the speed of light and this may cause it to "suffer" double bending. I have not delved further into the real reasons behind this doubling which are possibly beyond me but it would be an interesting question. But it seems as though such a doubling is not part of any other calculations for other G.R. gravity situations or is it? It is also unclear whether this correction was made before or after the experimental measurements.

Now if gravity is caused by light, as proposed by my new theory (TOLG), then the bending of light by gravity (another type of light) must be seriously questioned. Current light theory doesn't seem to support any such interaction between light beams or photons from entirely unrelated sources. When photons bounce off each other in interference situations there is some connection or correlation between them. But general interaction between supposedly uncorrelated photons would seem problematic. Photons have no longitudinal dimensions and perhaps can never "see" each other. So what does this mean to one of the greatest "proofs" of G.R?

I don't challenge the observation of light bending near the sun made by Eddington and many subsequent results. However I do challenge the interpretation of these results. Interestingly, recent observations seem to show a more significant bending than that previously observed. Perhaps this is due to improvements in experimental methods and equipment or perhaps it is due to other factors such as changes in the suns atmosphere. Another observation that also seems a bit unusual is that the amount of bending seems to have large variations depending on the time and place of measurement. That is, when the location of the background starlight being bent is near different parts of the sun's surface, but still at the same angular distance from the sun, it is bent by different amounts. Surely this would not be expected if gravity caused bending. So what other possible causes are there for such light bending?

I have another much simpler explanation for the bending of light near the sun. It is based on established light bending physics and on the properties of the media involved. It is well known that the speed of light is not fixed but is determined by two factors, the permittivity (ε) and permeability (μ) of the media or region through which it travels. The formula for the speed of light is; $c = 1/\sqrt{(\varepsilon * \mu)}$ so any change in these parameters will change the speed of light. In particular an increase will result in a slowing of light travelling through that media. This effect occurs in water where the speed of light is reduced or light is effectively slowed down. Now as any high school student knows, if light is slowed down through a media it causes bending at the interface. So any change (increase) in these two parameters, especially as light passes through the media, will obviously cause a speed change (slowing) and this will cause light bending.

It is well known that the sun has an "atmosphere" that extends several millions of Km into space. This atmosphere is not a conventional one but is made up of plasma or hot ionised gases. It is also to be expected that hot plasma has different electric and magnetic properties compared to free space. In particular the suns atmosphere (plasma) could influence both of these parameters significantly or at least the electric permittivity. This would result in a (significant) change in the speed of light in this region compared to free space. As explained before speed change is the basis of refraction or light bending. Therefore speed change and refraction through the suns atmosphere and hence near the sun causes light bending. So simple!

So the "atmosphere" surrounding the sun bends light passing through it. The next question is; by how much? I have yet to estimate a likely change in these parameters and the amount of light bending that would be produced, but believe it would be enough to provide a complete explanation for the bending observed? It surprises me that refractive bending due to media effects doesn't seem to have been considered as a possible cause of any bending of light near the sun. At least I have not found any reference to this possibility in my research. Yes I know there have been other earth bound G.R. gravity light "bending" experiments by well-meaning physicists that have provided further "proof". But I believe that the "jury is still out" on many of them at best. There may be other simpler reasons for the results of these experiments.

But what about the other related observations of light bending by gravity such as so called gravity lensing. Again, it definitely seems to occur and I don't challenge the observation, just the interpretation. Is it possible that the bending of light around "active" bodies (suns) is also due to an induced speed effect? Can the lensing around large cosmic bodies be explained by light speed changes due to plasma or other "gas-atmosphere" effects? I am not sure if this could cause sufficient bending and hence fully explain lensing in a similar way to how it is "fully" explained by G.R., but perhaps it could. Are there any other possible causes such as high level E/M fields that could provide sufficient light bending? Perhaps other possible causes for these observations need to be considered before we automatically "blame" G.R. and gravity.

QUANTUM PHYSICS

This work and my new ideas should have a significant impact on quantum physics. They may not challenge all existing quantum theories or models but will surely add new dimensions to many of them. And while I have proposed a major revision to how the proton is modelled, I have yet to develop a full or even partial set of new quantum parameters or numbers for the dynamic proton. Perhaps the major challenge is to the quark model which proposes that a proton is divisible into subcomponents. These quarks have many quantum properties including in some cases fractional charge but they are supposedly indivisible. I strongly doubt the correctness of the complete quark theory and believe the proton may be as indivisible and fundamental as the electron. I also strongly believe that the two particles, the electron and the proton somehow combine under special circumstance yet to be defined, discovered or understood to form a neutron, or vice versa, result from neutron breakup. Another bold theory and more work.

ATOMIC FORCES AND HEAVY LIGHT

Do my ideas have any relevance to atomic forces? Well, perhaps not as obviously as for cosmic forces such as gravity. There are forces at work at atomic level that make the atom what it is. The main forces are supposedly the strong nuclear force, the electroweak nuclear force (combination of electro (coulomb?) and weak nuclear force) and perhaps the force of gravity. But what particles does the force of gravity apply to and at what subatomic distances? What forces keep electrons and protons apart but "connected" in some form of mutual orbit in an atom? And what forces apply to neutrons, if any? Are they related to the possibility that neutrons may be made up of an electron and a proton and may be polarised? And how do all these forces relate to quantum properties of particles and atomic behaviour like atomic bonding?

Both the electron and the proton are charged particles that have identical but opposite polarities but significant mass differences. Does both gravity and the electroweak force apply or do simple opposite charge attraction or Coulomb forces keep them in mutual orbits? And does gravity apply only to all nucleons (hadrons) or is it just protons? In that case what keeps neutrons together? Are they kept together by the strong nuclear force that supposedly only acts over extremely short subatomic distances or does some form of localised Coulomb force apply due to their polarisation? Obviously I am no expert on these questions but the current theory about these forces and how they work may need to be reviewed in light of my ideas. Heavy light and TOLG gravity may have a role to play with protons and perhaps all hadrons but not electrons.

A related question is how far into the atom does PP light penetrate and what particle size and mass does it interact with? Is atomic gravity only caused by CGBR or is there other PP light at work? Does PP collision occur on the complete nucleus or on all hadrons or just protons? Hence on what subatomic particles does PP momentum and pressure apply? This question is related to the "size" of photons (PP) and is discussed in the chapter on new light. It is proposed that CGBR impacts on all the larger subatomic particles or nucleons such as protons and neutrons but not electrons. But this area is not well understood, at least by me, and needs more consideration. Hopefully this will come from a comprehensive analysis of and a mathematical solution to my ideas and from work on the "capture" and "measurement" of new PP light.

What role does PP play in atomic radio-activity and is PP light connected in anyway with atomic decay? Is the process of PP collision fully elastic especially for large "heavy" atoms or is there some absorption. Is this process of PP capture associated with some types of "spontaneous" radioactive decay (and hence fission)? Elements with high atomic number and a denser nucleus with more nucleons may possibly be more susceptible to attack and break up from background PP. There may also be (temporary?) capture of some small amount of PP energy. The CGBR gravity theory proposes that elements generally do not absorb energy from CGBR. The reason for this needs more analysis but is related to the frequency and energy of background PP (CGBR) which may be too short in wavelength to be "seen". Perhaps protons can't absorb PP when they are in a stable atomic state and therefore reflect most PP light. But occasionally the right wavelength PP may come along and be captured causing nuclear activity and radioactivity.

Therefore it seems that PP at low energy levels, especially gravity photons or CGBR, is not (easily) transferred to atoms and they just bounce around in elastic collisions. Therefore, they only impart momentum, not energy, so matter doesn't heat up from the impact of gravity photons! So CGBR could be said to be cold. That is, the frequency and hence energy level of a gravity photon is so that it can't be readily absorbed at atomic level. PP can't "see" electron clouds and they just pass through and only sometimes see a nucleus. They interact at nuclear level and only impart momentum at this level. But the nucleus is relatively small even with orbiting protons and the so called capture area is very low compared to overall atomic size and atomic spacing. The ratio of capture areas has implications on the probability of collision and is the reason why most gravity photons have a high degree of penetration. The rate of PP required to cause the necessary gravity pressure must allow for this high penetration factor.

Of course as one mystery of the universe is "solved" more questions are raised. Even if PP is discovered what does this mean for CGBR? In the case of gravity due to light "pressure", the obvious question is; what is the source of the new light? It is most likely to be from some very high temperature, high energy cosmic sources. The energy required to produce such large amounts of PP must be very significant and perhaps way beyond what happens in our solar system. A PP generator must be much hotter than our sun. Our sun is a good EP generator but perhaps not a very good PP generator if at all. If it is, even slightly, this has implications on the suns mass calculated from gravity forces. There are many potential very hot cosmic bodies but how do they produce PP? Or was it all created in the Big Bang just like CMBR supposedly was? Wow, I hope not as I don't believe in the Big Bang as a cause of CMBR or of anything!

GEOLOGICAL IMPACTS

The earth moves in mysterious ways under internal geographic forces and external forces such as gravity. The two main types of movement are complete movement as a rigid body (external movement as I call it) and movement which changes a bodies shape or causes distortion (internal movement). External movement includes all orbital motion as well as any large scale cosmic motion. The earth is perhaps moving at considerable speed due to all of these effects. If BBCM is correct and as a Copernican the earth is very typical of cosmic bodies, this movement must be at a very high speed. If BBCM is not correct, it would be a lot less.

The earth's internal movement are perhaps primarily due to thermal and rotation effects causing the core and surrounding molten matter and the earth's crust to move. These cause shorter term tectonic plate movements and longer term continental drifts over millions of years. Internal movement causes earth quakes and volcanoes and other effects such as high speed Tsunamis but perhaps not extinctions. These seem to be caused primarily by meteors and man.

But are all such internal movements only caused by internal factors or could transient variations in external forces over time and space also cause them? There are already regular short term variations in gravity forces over and above the long term average gravity pressure that keeps earth in orbit. Some of these occur at different times and at different places and some are more global and regular causing the tides. But there may also be other more transient variations in CGBR intensity at various locations on earth over time on a short or longer term basis. These may be sufficient to cause short term variations in external gravity forces (real gravity waves!) which may then cause transient internal earth movement.

Such variations may only be very minor and only last a few seconds or even a few days but over time they could have an impact. They could change the earths shape slightly and be sufficient enough to cause earthquakes and other geological activity. The likely causes of such variation are possible connected with the method of creating CGBR. They could be associated with many large scale cosmic events such as super nova explosions. Perhaps this is really what gravity waves are all about! And when CGBR is capable of being detected and measured this may provide a method of real gravity wave detection and hence some earthquake warning.

ANY DOUBTS

The most challenging area of most of my new ideas is that they are based on a specific light creation process and model of the photon. It is assumed that a photon is not created by the combined or composite action of an electron/proton pair. Creation of photons is assumed to be based on the isolated action of each type of charged particle. Each EP is created solely by electron activity while a PP is created solely from proton activity. My photon model also assumes each photon is produced instantaneously and without any connection with any other photon, even if from the same source atom/electron. In particular each photon is related to a specific atomic particle energy transition. Photons are finite size flat discs of fixed energy with no longitudinal dimensions. Photon size is related to wavelength which is related to energy. At this stage these assumptions seem consistent with existing light physics and atomic theory. But if they are incorrect many of my light assumptions and derived theories become challenged.

But if photons take time to be created, albeit of very small order (e.g. < E-10 seconds) then they may have some form of longitudinal dimension. Or the use of Special Relativity and Lorenz transforms to prove they can have no longitudinal dimensions may not apply at light speed and hence to photons. If photons do have a longitudinal nature this together with motion may explain red and blue shift. On the other hand, even if the photon creation process is effectively instantaneous, there may be some form of multiple sub photon creation process. Photons may be created in groups or bursts in a synchronous way with a fixed time separation. This may be related to a form of atomic/particle oscillation during the (multiple) photon creation process.

Photon wavelength and observations such as the photo-electric effect and atomic line spectra may then be due to the harmonic nature of multiple synchronous photons in each burst or group. Each burst may contain the energy according to Planks law, not each individual sub photon. This harmonic nature of bursts of photons could also explain many of the strange behaviours including spectral line width and red shift.

The separation and hence "wavelength" of each group would be changed by the motion of the source relative to the direction of release of the photon burst. And at capture the motion of the capture device relative to the motion of the burst could effectively change the separation and hence wavelength of the burst of photons. The motion toward the light direction would shorten the wavelength producing a blue shift effect while motion away would lengthen it producing red shift. I only came up with this burst photon model after I had almost completed the book. I decided to proceed with what I had already written instead of redoing it based on this new model.

In fact this new idea raises many more challenges about what a photon is. Is a photon the discrete sub packets within a burst or is it the complete burst of energy? And how long is the burst, what determines how many "individual" components are in it and what each component contains? How does this burst model of a photon then relate to the atomic/electron energy level jumps that create or capture it? And how is energy allocated across the multiple sub particles? This new model may provide more answers but it also raises many more questions. Unfortunately I must apply my usual escape plan in that I don't have the time, skills or resources to follow up on this new photon model. But I believe it is worth consideration.

FURTHER WORK

All new theories in this book and my TOLG book are based on sound scientific ideas and methods. But just as I have said before they are just a theories, open to criticism and review as long as it is scientific. They should not simply be rejected simply by saying "who are you to challenge the greats"? While the new theories of light and gravity produce some outcomes that may not yet be readily testable, it is not meta-physics. They should eventually be generally provable and testable, although to do this may require new technology and new mathematics.

There is an urgent need to search for the new type of light which is the basis for my new gravity theory. The main task will be in proving the proton model and finding and identifying possible sources of this new light. Research using high energy particle accelerators is an ideal area to start. Possible ways of creating and detecting PP in these environments need to be devised. Another area is to take a new look at cosmic experiments that are happening all the time. New interpretations of old data may be useful in this regard. Solar light bending and gravity lensing are good areas to start. Work to develop the new field of physics based on the new dynamic proton should also start as soon as possible. It will also lead to new discoveries. Whatever happens, I am sure my new theories will stand the tests of scientific scrutiny. I am sure PP will eventually be provable, creatable and measureable and become useful new tools in physics.

THE LAST WORD

When I first started to write this book I was uncertain if I would ever finish it and at many stages almost gave up. So here I am near the end of my second book. Perhaps it is not really finished as it seems to have produced more questions than answers. Yes I can hear readers say they never thought they would get this far either. Well let me say thanks for staying the distance. Did you learn anything? Perhaps not but at least you have experienced some new ideas about light. Yes, at this stage they are only ideas or basic theories or should I say proposals as to what may cause some strange light behaviour. They need more work, especially with the possible new model of a photon being a burst of sub-photons.

Lots of high level effort will be required to finalise, prove and apply my ideas and make then useful and possibly derive benefits from them. Perhaps this is all still a long way off but hopefully this book will stimulate some initial studies. These will enable the physics of light to move forward in new directions as a result. Physics hasn't had such an opportunity for a shake up since Einstein was a young man.

As you will have noted during reading this book, I have regularly said this or that area needs more work. Well I was not just saying that to fill the book although that is what may have happened. In a way I was asking for help. Perhaps not too subtly but nonetheless I need it. I am not a Professor of physics or even a postgraduate physics student studying for a doctorate (I wish). I am merely a backyard physicist with few bold ideas. I have very limited resources, not much time and am a bit challenged in the areas of specialist knowledge in physics and mathematics. But I still have a drive to research physics theories, challenge them where I believe they are deficient and promote my bold ideas. I will carry on and hope that others with the required expertise will pick up the challenge and provide input, mostly positive but possibly mostly negative! All sound scientific contributions are welcome.

But before you put the book down, this is not the end. To complete the book I have added two more chapters. They are not specifically on light and don't contain any bold new physics theories. They just contain some ideas I had on physics, its laws and process and on the future for physics (not of physics). Unfortunately they may seem a bit rambling towards the end as I have used them as a kind of thought dump but that is writer's privilege. They don't provide any Bacon like truths or fundamental and universal strategies for conducting better scientific research but should help progress science in a limited way. I am sure they relate to my aims of trying to improve benefits for mankind but perhaps in an indirect way. I am sure you will agree.

CHAPTER 8 - LIES (LAWS) AND STATISTICS

BIG BUSINESS PHYSICS

Physics and perhaps all science in general, is now Big Business. There are CEOs, Business Models, Marketing Plans, Schedules and Budgets to name a few of the similarities with the world of big business. Perhaps it needs to go that way as the resources, people and costs involved are now enormous. Long gone are the days of individual (mad) scientists in small dingy labs with limited low cost or even no technology, producing profound theories, many of which never saw the light of day for years. These days, physics seems to be on a massive scale, involving huge costs, the latest and greatest technology and global collaboration by many participants. New observations, even the smallest, often receive instant "publicity" and usually become sufficient evidence to support some strange new theory. There is a mad race to be the first to market (to publish) even to the point of sometimes "cooking the books" or at least gilding the lily with "stretched interpretations". Again, just like big business is doing all the time.

But even with today's extensive knowledge of science and a vast array of interlocking science theories, there is always room for new ideas. There are still many questions unanswered and while new ideas may answer existing questions, they often pose new problems. As fast as questions are addressed new ones arise and many important questions remain unanswered for years. But while some new ideas in physics seem to get exposure, some get little if any attention, especially those from outside the established scientific community. Perhaps the most important ideas are the ones that challenge existing theories and propose big changes. But these are often subject to severe criticism, struggle to get any objective scrutiny and usually don't survive. So how can bold new ideas generated outside, be brought to the surface, at least momentarily?

Obviously there needs to be some process that weeds out scientific scams and other crazies such as perpetual motion machines that generate free energy. But bold and seemingly far-fetched ideas are not always necessarily wrong. Even bizarre and unsuccessful ones may help to redefine the problem or reshape the boundaries. Some "clear the deck" for better ones just like the idea of an ether for light waves. Some new ideas may initially be primitive and require modification or continued development. Some may not survive initial scrutiny by "experts" at the time but resurface later to great acclaim under new science. Is there a need for a review and perhaps a change in some of our processes for acceptance and publication of any new ideas? Is the peer review process foolproof? How well are the popular science magazines doing the job of providing an entry point for new ideas from the public? And what about the internet?

One fascinating thing about encouraging scientific endeavour and promoting new ideas is the capability that the internet and a more open scientific world could provide. The internet is global, easily available (in most counties) and easy to use even by children. It may possibly provide a path for new ideas and theories to rapidly emerge and get exposure. But the internet has problems in exposing any new ideas. Good ideas can quickly get lost in the noise of junk, punk popular low brow entertainment and commercialism that the internet is famous for. Perhaps a new special forum like Wikipedia is required. But one which allows more open discussion. I tried to submit my ideas of light gravity and proton photons to such forums but they were almost immediately deleted by overzealous "experts". Perhaps I need to keep on trying.

SCIENTIFIC MISTAKES

Scientists make "mistakes"; they (we) are only human after all. Many books have been written in recent times about scientific mistakes. While they are less likely these days, they still occur and are usually more serious. Scientists (experts) often make errors or mistakes outside their area of expertise. This may occur when the media and public are fooled into believing an expert in one field is an expert other fields. While this may be true in some special situations, it is the exception rather than the rule. But unfortunately mistakes are also made by scientists within their area of expertise. Many are due to bad science but some are due to ignorance or even arrogance. Most mistakes are unintentional but unfortunately many are made intentionally, either for career protection, advancement or more often for financial gain. Many so called "discoveries" are often timed to encourage ongoing funding. Some "mistakes" lead nowhere, some are proven incorrect, most are quietly buried. Usually there is no refund for physics failures.

So it should be recognised that just because an "expert" says something is "true" doesn't mean it is. I have a saying that goes a bit like "If I had a dollar for every time an expert made a mistake I would be rich". But why would I make such a comment. Without a scientific background comments about science mistakes may not be well founded and would not be taken seriously. Perhaps I am not a PhD graduate nor Oxford Professor, but I do know basic science and often know a scientific mistake when I see one. But who knows, perhaps even I would support perpetual motion if it meant financial gain. Unfortunately, I am not yet famous enough to get any such offers. Don't get me wrong, there may be value in making some types of "mistakes" in science. Perhaps the error in "trial and error" is an important part of the scientific process. But unfortunately errors, wilful or commercial or otherwise, have caused too many problems to ignore. Therefore all errors should be identified and openly corrected as soon as possible.

However, the purpose of this chapter is not to dwell on science mistakes or to bury science so much as to (p)raise it above such problems. It is a bit of a clique but the search for scientific truth must continue against all opposition, even that created by science itself. And it must always be remembered that science theory is just that, theory. It is the "best currently supported" (most popular?) theory but is not "fact" or "true" and perhaps never will be. Scientists and everyone for that matter should continually be challenging scientific theory, scientific processes and so called laws of nature. Not to the point of denying everything and blatantly ignoring the advances and benefits science has produced like some luddites do, but to constantly challenge almost everything within reason. This is perhaps the best way to reduce errors, prevent mistakes and improve and advance science. It is a necessary part of the scientific process.

There must surely also be room for improvement in the scientific processes? How solid are the so called foundations of scientific theory? Are all current theories the best or just the best supported? Do they need regular servicing or health checks? It seems unlikely that current scientific theories and models of reality are profoundly wrong because they all seem to fit so well together. But we should never forget the geo-centric model of the universe. It was wrong but was continually refined with epi-cycles in order to solve any observed anomalies and keep it in favour. At that time and even up until quite recently, challenge to existing (science) theory was quite severely dealt with. The old establishment often overzealously guarded its "jewels".

Some scientific theories are said to be too big to fail. They contain such fundamental truths that they seem to be impervious to change and go on forever. Perhaps it is more the case that some scientific egos are too big. While there may be benefit in supporting and retaining sound scientific theory that fits existing observations and other theories, there is also the potential that unsound or even incorrect theories may be perpetuated beyond their use by date. Perhaps the inertia of existing theories, laws and processes as well as the ego of those within the ivory towers slow down the process of change unnecessarily. In many areas, old ideas seem to become so entrenched that critical analysis and necessary review almost becomes impossible. Is the new scientific establishment also over guarding its "jewels"?

Perhaps all theories age and need replacing by younger stronger ones just like in real life. Most scientific theories are found wanting in due course and are superseded by more complete or more comprehensive ones that apply to more situations. In many cases old ideas and theories have served their purpose. Progress was made because of them but eventually it becomes time to move on. New theories may be more complex but are also often simpler with "fewer epicycles". New ideas are always needed. Often they lead nowhere but may have merit and go on to help improve knowledge and progress our understanding of the world around us. In some rare cases old, previously discarded ideas are resurrected, reinterpreted and lead on to better science. In others, bold ideas proposed by outsiders, initially rejected by the establishment, go on to make significant contributions. However, if a new idea proposed by an outsider challenges existing theories or so called standard models, few experts are willing to consider it let alone investigate it or even support it. Let's hope that doesn't happen to my brave new ideas.

One of the real problems with new ideas or theories is not the generation process but the proof and acceptance processes. To get any acceptance theories must have some initial scientific justification or "proof". New theories must undergo some form of acceptance testing and "proof" by the initiator. Then they undergo more detailed scrutiny in a formalised peer review and acceptance process. Detailed analysis, further experimentation and objective testing are carried out by a broader expert community. All of these are key aspects of the modern scientific acceptance process. They have helped to ensure the development and acceptance of sound scientific theory. Trying to prove a theory or, in most cases, trying to disprove it, is what progresses science. But any thorough proof or "disproof" requires significant resources, expertise, effort and cost, mostly only available to the closed scientific community. It seems as though there is very limited entry for outsiders to these resources.

So what is the basis of an acceptable proof? Firstly there is a problem for non-experts like me to conduct thorough scientific and theoretical analysis of a new idea and prove it is correct. However, if it can't be proven to be correct at this early stage doesn't automatically mean it is wrong. I believe my bold ideas on light have a sound scientific basis. They now require acceptance as good ideas, closer examination by the scientific community and be subject to a better proof process? I am sure they will survive such scrutiny. Perhaps this book will set in train a process that will eventually prove my theories are correct, perhaps not. But in the meantime, if anyone can directly prove any of my ideas are incorrect, and that really means proof beyond scientific doubt, then I will surely listen. But perhaps even that will have difficulty getting support. Hopefully not too many dis-proofs will be found but I will be interested in any views.

Many theories have been proven or at least better understood by "accident". The photographic plate that was exposed when it was inadvertently placed near a sample of radioactive material is an example. This unexpected situation improved the understanding of radioactivity. Many other unusual outcomes or observations were interpreted in surprising ways. Perhaps the interpretation of an event rather than the event itself is the most important outcome. There are many events happening all the time that are awaiting the correct or at least a better interpretation. So what events or observations can be reinterpreted to help support or prove some of these new light theories? And what controlled experiment could be made to happen in such a way as to support my new ideas and help improve our understanding of light? What current situations or behaviours of light can be re-examined in an attempt to find a better explanation?

Of equal importance to a review of science processes is a review of the basic "truths" behind many if not all scientific theories. In particular, areas that should be examined are the so called basic "laws" of physics, the so called universal "constants" of physics and the "processes" of physics. One of the key components of the laws of physics is the conservation laws. These are so basic to physics that they should be continually reviewed, not just by quantum physicists who constantly break them to support their theories but also by physicist in general. It would seem that there should always be a search for a more fundamental basis for these so called immutable laws and any what if any limitations are behind them. Another key area is the seemingly ever increasing list of physical constants (and quantum numbers) that are necessary for most theories. Where do they come from and are they all really constant and necessary?

Constants that have been derived or discovered play a key role in almost all scientific theories. But are they absolute? Do they have the same value over all space and time or could they vary in different situations? Are there more to discover and are some related to others in a way yet to be discovered? The so called fundamental constants of physics such as Planks constant "h", and the gravity constant "G" (first derived by Newton then used by Einstein) and many more should be constantly (no pun intended) checked. Is "G" really fundamental or is it a function of other more fundamental constants as per my gravity theory? What other relationships exist between constants? Is there a finite small set of quantised, universal constants from which all others can be derived or are they open ended? How many if any are completely relativistic and how many need to be adjusted for relativistic (light) speeds to be compatible with quantum theory? And how does quantisation and Plank limits fit in with constants?

Mathematics has also become an important foundation to almost all our current scientific knowledge and science theories. New mathematical methods, like calculus, were discovered (created) to help solve dynamic motion problems of physics. Seemingly pure mathematical methods have found applications in new fields of physics such as matrices and group theory. What if a completely new mathematical concept was discovered? This may seem far-fetched or very unlikely, just like many new scientific discoveries made so far, but it is not impossible. So just like all other theories of science, the application of mathematics to science need to be constantly evaluated. If a weakness was found or improvements made in some parts of existing mathematics what impact would this have on the theories based on it? They may be significantly and even negatively impacted and this may have serious implications on our scientific world.

During my work on gravity and light, I came up with some basic questions about the application of maths to physics problems that challenged me. It was not just that the maths challenged me, although that is certainly true of some aspects of theoretical physics, but particular outcomes also challenged me on many occasions. The use of mathematics in theoretical physics should be constantly challenged. Some of the challenges in the application of mathematics to light and gravity theories are addressed in other chapters of this book and in my earlier book on gravity. But some needed more discussion and analysis so I have included them here. I believe the issues relate to some of the fundamental questions in physics that mathematics has played a key role in creating and "solving". Are the solutions and the mathematics behind them robust and reliable?

While I would like to investigate what is behind the general laws and constants of physics, especially the related concept of Plank limits on some variables, these are beyond this book. And I am not in a position to create a new mathematical process to solve some new problems such as TOLG. Instead I will review the two areas of physics that have fascinated me most of all and I believe need reviewing. These are the conservation laws and the use of mathematical equivalence in physics. Is there a fundamental basis for conservation laws or are they merely an outcome of the development of science? What is the real meaning or reason for them and what if any restrictions apply to their application? What about the role of mathematics in physics? What limitations or boundaries if any, apply to the relationship between these disciplines? Perhaps there are none, or at least none that have been found yet, but will this always be the case?

So let's examines these two areas, the basic conservation law(s) and the role of mathematics in physics. I will attempt to identify some of the issues and hopefully answer some questions or at least formalise and define them a bit more. This is small but positive progress I believe.

THE LAWS (LIES?) OF PHYSICS

In physics as in many analytical or scientific disciplines, there are a few so called laws that seem basic to everything. They apply to all existing theories and any new theory or idea must satisfy them in order to be accepted. These laws of physics have stood the test of time and proven to be invaluable in helping to prove or disprove theories and progress science. Non-compliance with these laws has often been the downfall of many brave new ideas. Compliance doesn't guarantee success but is a necessary component for it. But in recent times some unusual exceptions to these laws of physics have supposedly arisen, especially in the new areas of quantum physics, atomic physics and cosmic physics. In very special cases some of these laws must be violated for the new theory to be accepted. Or does the theory perhaps have a fundamental weakness.

One of the laws of physics that has played such an important role in the success of many theories is the conservation law or should I say set of conservation laws. The basic law is that some physical property such as energy is conserved and cannot be created or destroyed. It can only be converted from one form to another. A special addition to this law was derived in Einstein's Special Relativity which stated that mass or matter is just another form of energy. This was used in the development of nuclear physics but I am not sure how reversible the process is or if energy has ever been converted to matter yet, either in a controlled or uncontrolled way.

In general, conservation extends to many properties of matter some involving motion and some involving fields. A number of physical properties of matter and movement seem to be conserved in many if not all situations. Certain properties of a body remain conserved independent of any external or internal change that may occur to that body in space and time. For example motion energy or kinetic energy of a body can be converted into potential energy in a field. In the absence of a field, motion energy is conserved unless converted by a (non elastic) collision into heat or sound energy (another form of heat energy?) or both.

A simple set of laws that seem to apply to all matter include;

1. Conservation of mass
2. Conservation of momentum (linear and angular)
3. Conservation of energy (linear and angular)

For a given body of mass **M** and for properties of length **L** and time **T**, there is a common law for the conservation property that I have called **CONV.** These laws all take on a basic mathematical form given by:

$$\mathbf{CONV(n)} = \mathbf{M * (L / T)^{n}}$$

This value CONV (n) always seems to be conserved for various (integer?) values of the index n. For example when n is zero, 1 or 2 this produces three well known laws.

For n = 0 the law becomes

$$\mathbf{CONV(0)} = \mathbf{M * (L/T)^{0}}$$
$$= \mathbf{M * 1}$$
$$= \mathbf{M}$$

This is the conservation of mass law.

I remember using it in experiments involving burning material and measuring the before and after masses and making assumptions based on mass conservation. Does this law always apply or when if ever does it break down? It seems to break down in some atomic processes and at relativistic speeds when mass and energy (in some other form) seem to become alternative forms of the same thing according to Special Relativity!

The next case is when n = 1 and where V (velocity) = L/T, the law becomes.

$$\mathbf{CONV(1)} = \mathbf{M * (L/T)^{1}}$$
$$= \mathbf{M * V}$$

This is the conservation of momentum law.

But does this law always apply and if not then when does this law break down? Does it also break down at relativistic speeds when mass and energy (in some other form) seem to become inseparable? This is an interesting question but is outside my area of expertise.

The case of conservation of angular momentum is for further study but should provide some interesting outcomes. The length parameter is replaced by the angle and the ratio with time becomes angular velocity. But how are angular and linear momentum related?

The next case is for n = 2 and the law becomes.

$$\mathbf{CONV(2)} = \mathbf{M * (L/T)^{2}}$$
$$= \mathbf{M * V^{2}}$$

This is along the lines of a conservation of energy law.

Kinetic energy is usually expressed as; $\mathbf{E = M * V^{2} /2}$.

This is proportional to the conservation law above. But perhaps in a way the full conservation of energy law expresses a total conservation of a combination of kinetic and potential energies. Perhaps a similar amount of potential energy in any general field situation is required for the concept to apply. In any case proportionality applies. And when does this conservation of energy law break down? Perhaps it is at relativistic speeds again.

Does it apply to all subatomic situations and all cosmic situations for all times? Does it apply in G.R? And how does the conservation of angular energy fit with this?

A COMMON LAW

$$\mathrm{CONV}(n) = M * (L / T)^{n} = M * (V)^{n}$$

Conservation laws seem to have a common fundamental basis and can be covered by a single conservation law relating parameters of mass length and time or more specifically mass and velocity. The value CONV (n) is always conserved for some integer values of n (zero, 1 and 2).

What is the basis for this fundamental relationship between these three parameters? Does this law apply for any other powers of n? Perhaps the most interesting aspect of these conservation laws is that length and time by themselves are not important, only their ratio or the velocity. Is there something fundamental about the ratio of these two factors even at various powers of n? Is there something about conservation laws that implies that space and time are always connected?

Does it imply that time cannot be treated as another Cartesian space variable to make four dimensional space time but must always be connected with them as a ratio? I tried to analyse this question using Special Relativity. I wanted to see if the famous Lorenz transforms of space and time when connected in such a ratio shed any light on this question but to no avail. At this stage this is as far as I have progressed this idea or question but it seems to have more potential.

MATHEMATICS IN PHYSICS

Another challenge in physics is the suitability of mathematical processes used to derive, and prove theories. Physics and mathematics have evolved closely hand in hand since Newtons time if not before. Indeed this nexus between mathematics and physics has produced many great advances in science. They have evolved in parallel to a large part with each helping the other to new levels of complexity. The combination of mathematics and physics also led to the application of mathematics to technology and engineering, the field I entered. Mathematics works well with engineering and has produced a vast array of devices that have aided humanity in so many ways. It is why our very large bridges stay up and very heavy planes fly, at least in most cases.

But what brings maths and physics together? Is maths just a development of our thinking about physical problems or is it like an art form and often removed from reality. And what came first, mathematical answers (the egg) or physics questions (the chicken)? Newton developed a new form of mathematics (originally called fluxions, now called calculus) to help derive and prove his theories of motion, force and gravity. Fourier analysis is an example of classical mathematical methods that already existed and were then applied to physics. Einstein applied complex and somewhat esoteric mathematics called Tensors, developed by Riemann and Ricci, to produce his theory of General Relativity. Mathematics such as group theory and gauge theory which have been around for a while, have found new applications in the new physics. What other branches of pure mathematics, like Riemann's Zeta function, may come to the rescue again?

Today mathematics has become critical to the development of physics theories. Many new branches of physics are initially theoretical because of the difficulties in carrying out useful repeatable experiments or making practical observations. In some fields such as atomic or quantum physics as well as cosmic physics, research and experimentation is costly and difficult due to scale (facilities and objects that are too large or too small) and time (extremely short or very long event durations). In these areas initial progress is often made only with pure mathematics. A mathematical model is proposed, developed or formulated then used as the basis for further work. It is often the only way forward until technology catches up, if it can.

But theoretical physics is becoming very complicated and problems with mathematical complexity have been major factors in slowing down scientific progress. Some new physics, especially quantum theory and yes string theory, are almost entirely theoretical and based on complex mathematics. Many new theories are very esoteric and some, like multiple universes that supposedly arose out of string theory, are more like meta-physics and may never be proven or disproven. They are all mathematically "correct" though.

This relationship between maths and physics is very strong. It is so critical to the development of physics that without a sound mathematical basis, any new theory may struggle to emerge let alone survive. However, while maths is important, it may not be enough. It is necessary but not sufficient, as the saying goes. Eventually even a purely mathematical model or theory needs to be "proven" or at least validated and supported either by observation or experiment. New outcomes and predictions need to be made that can be observed or verified by experiment. And it needs to comply with basic laws and be consistent with existing models (or does it?).

Maths has a philosophical and perhaps even an artistic side while physics has a more practical or pragmatic side. But is mathematics always logical and "correct" or is pure mathematics just another form of theory that always needs checking. Can it be subject to improvement or even worse may it contain contradictions or errors that lead to erroneous physics theories? And are the fields of mathematics and physics completely independent or are they always closely related. If they are connected are they always a perfect fit? Is there any potential incompatibility between them and if so what would this mean for the theories of physics that have been developed based on mathematics? But before these types of questions can be addressed the two components of this discussion, physics and mathematics should be defined, at least in broad terms.

What is science, especially the specific branch of science called physics? In simple terms it is about trying to understand the how, why, when and where questions of nature and the universe from a physical point of view. It is about defining them, describing them and answering them by developing formal theories with specific parameters and yes making them measureable. But I will leave any further definition to the philosophers. So what then is mathematics? Is it science or is it art? And is it just about symbols and numbers, or is it more than that? I don't intend to delve into a rigorous definition of mathematics like Bertrand Russel but I will describe some of the basic concepts of math theory or is it math practise and explore what they may mean to physics. Sorry if this is not very rigorous but I am sure it will progress the discussion.

One very interesting aspect of mathematics is the concept of a zero. Many earlier number systems such as the Roman and Chinese systems did not have a zero. It was developed in the east in India or Arabia and eventually became a fundamental part of our current number system. But what does zero really mean? Does it mean nothing or the absence of anything or is it just a counting symbol or concept. And how is zero related to the concept of infinity or unboundedness? Zero is an important part of the number line but unlike other real numbers, its inverse is not on the number line. The use of zero causes problems in some areas of mathematics. Division by zero is not defined or can't be defined. And when zero is used in indices and logarithms, an assumption is applied to obtain a meaningful answer.

So given the potential problems with zero in maths, how is zero applied in physics? There are few situations where zero has any meaning in a physical sense. One situation where zero is applied is in measuring temperature. Absolute zero temperature is said to exist and is defined as 0 degrees Kelvin or -273 degrees centigrade. But the concept of an absolute zero temperature becomes difficult to define as the temperature scale at very low temperatures seems to become logarithmic. Absolute zero temperature has never been achieved and perhaps never will be. So does it really exist or is it just a convenient concept for our definition of temperature?

And how does zero relate to other physical properties such as mass, length and time? Is free space zero matter but not zero time? Free space supposedly contains zero matter and zero energy. The best example is outer space but even this contains a few atoms per cubic metre and of course some (many) random photons and perhaps even some dark things and fields. What is free space between and even within atoms? Is it nothing or is it probability fields? But even the idea of free space containing nothing is now challenged. According to quantum theory free space supposedly contains something or is it everything, at least from time to time.

Plank theory proposes that most parameters such as mass, energy, length may have lower quantised limits. Such parameters or measures are proposed as quantised multiples of basic non-zero building blocks. They cannot be reduced down below this basic Plank limit unit and therefore can't ever be zero! These limits therefore seem to restrict the possibility of a zero of anything for which Plank limits apply. But what about time? Is there a plank limit or can there be a zero of time like a concept of instantaneity? It seems not according to Special Relativity.

And does the concept of infinity, the mathematical inverse of zero, exist in physics? Is there infinity in either time or space and is the universe infinite? If so what does this really mean for the Big Bang theory? Is the only real scientific case of a zero in space and time the situation before the Big Bang or is it only the inverse of zero, or infinity, represented by the infinite universe? One of the biggest recent problems with the use of zero in physics (or was it infinity) was in quantum theories. A process called "re-normalisation" was used to make sense of some quantum theory mathematical difficulties when a zero prevented sensible solutions. However, the outcome may have been the creation of another questionable theory.

But the application of a mathematical zero to physics may not be the main problem in the relationship between the two. It seems to me that the main problem is the use of the most important component of mathematics; the equivalence concept. Equivalence is at the centre of all mathematics. It is the basis for relating numbers or functions of variables with other numbers or functions of variables. Mathematical processes based on maintaining equivalence are then used to reduce the number of variables and produce a solution or solutions to the equated functions. There are other expressions such as proportionality, approximately equivalent, less than (<) and greater than (>) and variations on these but they don't change the fundamental problem with the concept of equivalence. So what does equivalence in mathematics really mean for physics?

Mathematical formulae using the equivalence concept have helped progress science. Equivalence has both formalised and mechanised science and made it more rigorous and practical so to speak. But there is at least one fundamental problem with the application of equivalence based maths to physics. This problem was probably first realised by Newton in his work on gravity. It is not the problem of equating zeros or even approaching zero by using ever decreasing limits as in calculus. The "in the limit as a value goes to zero" concept was a major problem with calculus at the time and still is with high school students but is not a fundamental problem. The real problem is that mathematics doesn't seem to allow for causality or timing between equated events.

Physics formulae allow for time as a variable and many have time as a parameter. But this doesn't solve the problem. It only hides it or disguises it. In equations using equivalence, the two sides or events happen at the same time and for all times or at the times specified identically or simultaneously. Newton used the mathematical concept of equivalence to derive his (instantaneous) gravity theory. He directly equated force to the acceleration of the mass involved in has famous equation, F=m*a. In gravity he directly related force to the product of the two masses and the inverse square of the distance between them. There was no allowance for any causality or delay in the creation of the force or its action after it was applied.

Newton was fortunate because in that period, his laws were much more accurate than anything else. Cosmic measurements were still primitive and there was no real concept, let alone measurement of light speed. His laws provided an almost perfect fit to observations or reality as it was known about and measured in his time. There was no real understanding of the speed of gravity and how it may compare to the speed of light. Concepts such as Special Relativity and all it implies for the maximum speed of a body, energy and information were effectively "light years" away. All was quiet on the scientific front with physics and mathematics working well together using simple equivalence. But perhaps he realised there was a problem with an instant acting force, especially over a large distance. Later, many physicists, including Einstein, had problems with Newton's concept of an instant force of gravity and also with simultaneity.

As measurements became more accurate and physicists questioned the concept of simultaneity, problems soon arose. Questions about what really causes gravity and what is the speed of gravity were being raised. The speed of light and the concepts of relativity had to be factored in, but how. Einstein solved some of the problems associated with simultaneity with Special relativity and with the speed of gravity with General Relativity, but did he? His maths still used equivalence or the mathematical version of it and made no allowance for causality. His G.R. theory was based on a concept of the absolute and identical equivalence of force and acceleration. Was it a product of the mathematical concepts he used based on simple timeless non causal equivalence or was it a fundamentally new concept as proposed?

Another related development in quantum theory was the derivation of the uncertainty principle. This seems to suggest that in many cases, it is not possible to determine both time and space related measures independently. A specific property or metric of a body such as its location cannot be determined if its momentum (speed) is already accurately known. Only a probability (less than 1) of a particular value can be determined. This seems to occur when two properties such as momentum and position are closely related. If one is known accurately then the other can't be determined accurately. Not sure how this relates to causality or to the length to time ratio mentioned before but it may so I included it out of interest.

Current mathematics allows for any variable to be used in equations and many physics equations have time as a simple mathematical variable. In many mathematical methods used for physics such as vector calculus, the Lagrangian, the Hamiltonian and many more, time is just another variable. In mathematical terms a variable can usually be positive or negative or even zero. But what does this mean with time? Can time be negative, zero or even run backwards? The maths says so but in reality this seems to cause problems. I won't go into the many conundrums this creates such as going back in time and killing your parents before you are conceived, but there are many like this. Perhaps the problem is caused by simply including time as another general variable in mathematical equations. Perhaps simple equivalence doesn't really work in all situations especially with time as a variable!

Another example where time is used as another type of space dimension is four dimensional space-time. This four dimensional concept was developed to help explain relativity and has evolved and found many new applications. It has been extended with additional dimensions (all space?) to develop physics theories such as string theory and to try and solve the "unifying problem" but with limited success. But what are the limitations of this practice if any? Does the inclusion of time as another space like dimension solve the causality problem, I don't believe it does. Physical events and interactions are mostly not as "simultaneous" or equal as the equivalence concept indicates. Many are related through causality and are not simply identical in a simple mathematical sense. To properly address these situations may require a new type of mathematics, I believe. Allowance needs to be made for causality and a time delay relationship between linked events defined on either side of an equation.

So how can mathematics be "corrected" for causality? How can equations be modified to allow for causality and if possible the speed of causality? This problem was supposedly solved for gravity with General Relativity (GR). The speed of light was included in Einstein's gravity equations. Most problems that had been identified with Newtonian gravity such as the perihelion advance of Mercury were solved. But it still didn't seem to really allow for causality. In fact it seems that Einstein in developing his theory implied that causality is not a part of gravity. The so called equivalence principal upon which G.R. was founded seems to suggest that force and acceleration are identical (simultaneous?) and not causally related. I still have a fundamental issue with this assumption but perhaps that is my problem. So what is the solution? Well if I knew that I wouldn't be here writing this book, I would be famous! But a solution to this problem must be found.

This situation reminds me of the problem with the early quantum mechanics. The high speeds involved required the speed of light and Lorenz contractions somehow to be catered for. Quantum theory didn't work well until the jump was made to relativistic quantum theory. In particular from Schrodinger's wave equation to the Klein–Gordon equations. I won't go into details as they are a bit beyond me and this book but suffice to say, better results were obtained when quantum theory was corrected for special relativity. In engineering, time related causality problems are very common. Feedback problems in control theory are addressed using time as a parameter in complex mathematical equations. I remember using the "S" plane concept with poles and zeros in the analysis of control theory problems. But engineering system speeds are relatively slow so it worked satisfactorily in all conventional or non-relativistic situations. But is a relativistic version of this theory required for high (light) speed control situations?

Can relativistic speed based mathematics be developed as a substitute for causality in all situations? Is relativistic mathematics the only answer or is there a need for a new type of maths along the lines of how calculus emerged? How can time be included in a causal way without introducing time reversal conundrums? I don't have any answers but strongly believe this question is as important as the many questions about the theories of science we are facing today. I also believe it may be related to the question about the laws of physics set out above, in particular the importance of the ratio of time and space in the conservation laws. A solution could also provide a way forward for solving (or removing the need for) unification.

SUMMARY

I have discussed the laws of physics, especially the conservation laws and the use of mathematical equivalence in physics theories. How important are these two aspects of physics? Are there any limitations and is there anything more fundamental about them?

In the conservation laws, it seems that the property of most importance is a ratio of two key measures or physical values, length and time. The fact that conservation laws always use only the ratio may have some significance. Is there something fundamental about these two parameters always being together in a ratio but not as individual values? Does it suggest a relativistic or quantised (Plank limit?) adaptation is required for conservation laws?

Mathematical concept of equivalence may have implications on its suitability for physics. Equivalence only seems to allow for simultaneity or identical timing in all situations and doesn't seem to cater for causality. Does causality need to be included in mathematical equations or was Einstein correct in assuming causality is not an issue, at least in his theory of gravity. Are there other mathematical models in physics where such a problem seems to be limiting or preventing a solution from being found? My new gravity theory suggests it is a major issue. Time will tell if I am correct. Perhaps causality was not an issue for the old sub-light speed (non-relativistic) physics but most new physics is at or approaching light speed. And if causality does need to be included what new mathematics is required?

Unfortunately I have no solutions to either problem yet and perhaps I am not completely convinced that they are problems. Any ideas would be welcomed.

CHAPTER 9 – THE END IS NEAR

In line with the heading about the end being near, a religious view of the future or end of science may be appropriate. But this isn't about the end or the beginning of the end or even the end of the beginning of science (excuse the copied metaphor). Instead it is just about the future. Not about the future of science but the future for science and who or what will "shape" this future.

The history of science is well addressed in many good books and perhaps a bit in this book. The philosophy of science has also been the focus of much work but perhaps not as specifically as Bertrand Russell's work on the philosophy of mathematics. So perhaps this is an area for more study. But while the history and philosophy of science may be interesting subjects, looking forward seems more challenging to me. Is science becoming like a religion (if that is a meaningful concept) with all its dogmas, prophets and true believers? Science or at least physics should never really be like religion. It is not just based on something called belief but has many believers. It should be based on something called proof and be open to all new and even radical ideas unlike most if not all religions which are dogmatic and mostly closed to change.

But if science is not a religion then what is the relationship between these two fields of human endeavour? Is one more significant or more fundamental than the other? Does religion have any real significance beyond mankind or is it just another human creation? Is science only defined in terms of its significance to mankind and is it just another human creation or does it go beyond that? Are our physical theories and models just to help ***us*** understand ***our*** world or is science, or at least physics, really absolute and applicable to all time and space as currently proposed. I examined the logic behind some of the universal "laws" of physics and some aspects of the role of mathematics in physics in the last chapter but left questions unanswered. Are these aspects as universal as believed and can the absoluteness of both mathematics and physics laws be proven? And how will these questions and answers impact future science?

While I believe the basis of physics in a universal sense is an interesting area for future research, I don't intend to address it any further here. And I also don't intend to study or write anymore on the past history and philosophy of physics theories. Instead I think the soundness of our current knowledge of physics and in particular, how we can ensure its continued development are great areas to discuss. So perhaps the "religion" of science is more about the future role of science in society, where it is going and how it will get there and who the key players will be? I do not wish to address the future of religion, although that is also a very challenging question. But there are interesting parallels between these two fields of human endeavour. It may be useful to contrast these two fields to help look to the future for science or at least physics.

This comparison involves loosely-defined and broad issues but should still be useful and help make progress in some way or another. I am not sure how it will work out but here goes.

RELIGION VERSUS SCIENCE

Science and religion are related in many ways. Both were "created" by man in an attempt to try and answer the questions that have been raised by humans about life and the universe. These questions have been around for as long as we have been "thinking" and many are just as relevant today. Religion tries to explain the world, life and everything else using books by prophets, so called absolute laws and some beliefs like a super being and life after death. Science tries to explain everything and relate all aspects of nature to theories based on some beliefs in immutable laws and mathematical logic but not on a god or life after death, just a Big Bang.

Another branch of human endeavour that in a way connects science and religion is philosophy. Did philosophy evolve from religion or religion from philosophy? One view is that philosophy was an attempt to formalise religion but ended up creating something we now call science. Science supposedly evolved primarily from natural philosophy. But if science came from philosophy and philosophy came from religion, why did science become a separate field? Some say philosophy produces surreal answers; religion produces philosophical answers while science produces practical answers. But some scientific theories are just philosophical (meta-science) and some religious "laws" have proven to be quite practical. And where is philosophy now in this discussion? How important are philosophical ideas if not answers, where are they captured and who uses them?

So, of the branches of human endeavour, which came first, religion, philosophy or science? Did religion "establish" human civilisation and then did human civilisation create philosophy which then created science or was it the other way around? Did logical thinking and a logical, deductive, scientific approach create the tools and new ideas which started our civilisation and lead to modern man or was it religion that brought primitive man together as a "special team" and gave us the "desire" to improve our lives? Perhaps these fields of human endeavour were almost indistinguishable in early human history. They emerged gradually over a period of time more or less in parallel. Some say both religion and science were the same thing and only recently diverged into separate fields of human endeavour but others say they were always different. Others say religion came first and science arose later as a challenge to religion although that is also disputed. So when did they split apart or have they really separated?

Was if around the time of the Middle Ages when people like Bacon with his treatise on scientific process based on using experiment, observation and prediction, were questioning the role of the church and its dogma? Religion initially discouraged and even prevented such questioning, observation and learning. But it kept surfacing and became very strong in the Renaissance. Perhaps religion and science started to diverge when they started to disagree on the interpretation of new observations. Technology started to improve our ability to observe nature. More accurate and reliable scientific answers began to emerge about some of the oldest questions religion had already answered. But religious answers were found to be wanting or even wrong. A scientific approach started to answer more questions in a better way and also raise many new ones for which religion had no real answers.

What is the relationship or real difference between religion and science now? Both science and religion have and continue to face similar questions. Both try to produce "correct" answers so that they can stay relevant, be useful or popular and hence survive. Both are seeking to influence ideas and thinking and trying to control the efforts of man. As a result they have been competing against each other for common ground ever since they separated. They have had a complicated and difficult relationship because of this competition and this looks set to continue. Which if either will win or will they continue to agree to disagree when in conflict?

There are many parallels between these two fields of human endeavour but some important differences, especially in the areas of evolution. Scientific theories or laws are manmade but are also supposedly universal. We may have discovered them but they were supposedly there all the time. But what about religious laws, in what sense are they universal and beyond human involvement? And the concept of encouraging open review, challenge and change when justified, does not seem to apply to religious "laws" or doctrine. Are religious laws already perfect and completely immutable or beyond human challenge? And while science is expanding it seems that religion (or at least moderate non-fundamentalist religion) is declining. Will this divergence, competition and expanding/contracting trend continue or will religion become even more dogmatic and fundamental in order to survive a scientific onslaught? Where will it lead?

Religion is based primarily on supernatural forces and events, something called belief and of course a god(s), all of which can't easily be tied down by specifics. There have been many variations on religion over the ages and there are still many even today. Religion was originally based on the idea that there were many gods, one for each event, feeling, season or whatever. This started to converge to a new concept of a few gods based on cosmic bodies such as the sun and moon that were more mystical at the time but recognised as important to life. The idea of a single all-encompassing god, called monotheism, eventually emerged and now most "popular" religions are based on it. But some still have multiple gods for different occasions and even most modern ones still have an "anti-particle" or opposite to god called the devil. Yes, similarities with particle physics. But at least science doesn't have many versions or does it?

Religion uses a type of absolute, superhuman being or "perfect" person (god) as the source or creator of everything (including itself?). The image of god was originally based on a humanised male concept but I am not sure if "modern god" is still human like and male? Monotheism or the idea of an all-encompassing single god has been virtually unchanged since its inception thousands of years ago but even this concept is still ill defined. Perhaps this is because if a definition is attempted it is challenged and comes "unstuck". The concept or meaning of a god is as varied as there are people to ask about it. Every time an attempt is made to seek a clearer understanding of what god means, religion seems to oppose it or at the least confuse it with dogma or the emperor's new clothes "explanation".

Religious rules or laws supposedly came from this external and eternal being (god?). As such they are supposedly universal (in a cosmic sense). But do religious laws have any absolute basis or are they only human based? Most if not all religions seem to apply only to mankind. In fact it seems that religious laws are just made by man for man (and women?). They are only for here on earth or nearby space (including the moon and perhaps mars soon) and "heaven" wherever that is. But what about other life forms here on earth and life (human-like or otherwise) elsewhere?

Most religions also have a philosophy of an "afterlife". "Good" behaviour here on earth is said to lead to some form of good life in another (better?) non terrestrial environment. On the other hand "bad" behaviour leads to something a lot worse (in a land of anti-particles). Some religions follow this dogma to the point of encouraging self-destruction for "the cause" to get to the afterlife (good or bad?) quicker. This is now presenting a serious challenge to our safety.

While there are still some very primitive religions and some primitive religious beliefs even within Christianity, many religions have been "modernised" so to speak. Some have changed over time along with civilisation and continue to change or "evolve" as our understanding of "reality" improves? Modern religions have accepted scientific discoveries and knowledge. Some religions, especially Christian based, are now trying to merge by reducing the differences (our common god). But others are still very primitive and fundamentalist and fight against change. They are often anti-knowledge, usually anti-science, anti-secular and are strongly opposed to any acceptance or integration of other religions (the chosen people, my god not yours). A few religions are fighting to defeat their religious and non-religious "competitors" using violence and destructive methods reminiscent of the dark ages and even worse.

But even today, many basic religious ideas have not changed. Instead they are described in terms like " true believer" which are not very meaningful to many people and not readily definable. What does the word "belief" really mean? Belief is one of those ideas in our language that has a lot of so called historical baggage. It is a "soft", ill-defined term. It is not really a scientific term although it may be used occasionally by scientists to comment on some idea or proposal or theory or observation. Can the concept of belief be defined in any scientific sense at all? Science generally uses more soundly defined terms and of course mathematical equivalence and even probability theory, to explain events and ideas. But are these just another form of belief?

Perhaps the religious approach of using loosely defined concepts and ideas that can't be logically analysed or challenged is what may have kept it alive. But this may also lead to its downfall. If religion can't be pinned down by definitions, this may be what eventually marginalises it away. But perhaps it may survive because of its perceived value or believed benefit for civilisation like other more esoteric branches of human endeavour such as art, philosophy and psychology.

Religion in its many versions may be just a human creation. But is science anything more? What are the real differences between them? Questions arise such as: can science and religion really co-exist forever; does (something called) "god" really exist and did god create (the laws of) science and if so then why? And finally will religion evolve and survive or will it receded and eventually be replaced or superseded by science? These questions are at the centre of the debate about religion and science and also the future of both for society. Which, if either, is absolute, perfect or complete? Which has brought more value or benefit to our civilisation and which will continue to do so in future? Which is worth saving, either or both?

Science is based on definition, observation, experiment and measurement. Rigorous more formal processes such as mathematics and logical interpretation of observations are used to produce laws and theories. Science develops and uses technology and carries out repeatable experiments to produce measureable outcomes. The foundations of science are based on something called "proof". I put proof in inverted commas because it is a challenging concept. Perhaps it just means the theory that bests fits the available evidence. But it is still an important and fundamental part of the process. And science doesn't have many versions. There is usually something called a standard model which may have many uncertainties and questions but they are recognised as being subservient to the standard model. There may be non-standard models (such as some of my theories) but these are also not given much weight if any. Perhaps the most important point is that science theories are open to investigation, challenge and change. In fact while science theories seem fundamental and basic they are constantly being updated and replaced by better ones that often incorporate previous ones as special cases.

Perhaps another major difference is that religion is primarily about rules just for human existence while science is supposedly about rules for everything, including god? Religion was initially based on a human and earth centred universe. We were the (only/ chosen) people of god and hence we and earth were special and central in the universe. But science came up with some challenging ideas about the role of our planet in the universe, the origin of life and our importance to it. Copernicus proposed that our planet was not "special", Newton found coloured light was not special and Darwin reviewed our views on creation and evolution and questioned if human life was "special". Science began to imply (prove?) that nothing, even our sun, the earth, light behaviour and humans was special after all.

It seems that science continues to evolve while religion, especially some primitive religions, don't change and in fact strongly fight against any change. This is perhaps because followers of these religions believe they are already perfect and complete or are frightened by possible change! This unfortunate situation has and will continue to cause considerable human misery and suffering. Religion has supposedly been a key factor in civilising mankind. But it has also had a long history of bigotry, intolerance and causing human misery. Like religion, science has also gone hand in hand with our own evolution and has been a key factor in the emergence of modern civilisation. But science has also often been applied to unsatisfactory ends and in creating human misery. Perhaps both can be either useful or misused just like all "manmade tools".

To more and more people in many cultures, science has become the accepted method of trying to better understand and explain our natural world. Most Christians and some other religions have matured enough to accept science and come to terms with it. They have worked out how the two can co-exist to a certain extent, science in a practical sense and religious beliefs in a spiritual one, whatever that is. Fortunately, at least in most western societies, no one is "burned at the stake" these days (so to say) for proposing scientific theories or challenging religious dogma.

But science still plays a subservient role to religion for many people in some cultures. These people are indoctrinated into believing that (their) religion is everything and science is wrong and dangerous? Sometimes scientists are persecuted to the point of death by these religions simply for being scientists or proposing scientific theories. So will science become the major driving force for improving civilisation and eventually overcome all religious bigotry?

Science has undoubtedly extended our knowledge in so many fields of human endeavour and helped us to live better lives. Many say religion has done the same. But the role of religion in western society has been steadily reduced by the ongoing replacement of religious dogma by proven scientific theory. Like religion, science is also a human creation but the "laws" of science have evolved and will undoubtedly continue to do so, based on a scientific process. Religion is based on the idea of a god and "his" laws that are absolute. Science is based on the idea that the laws of science are absolute. Which if either is really absolute and beyond humanity?

Science theories are developed and used to model reality and more importantly to predict new outcomes. Scientific theories are based on a rigorous although not completely foolproof process of repeatable experiments, observations, research and analysis. These theories are then tested and the cycle continues so that more reliable, interconnected theories and models are slowly built up. These theories and models are used and applied to develop technologies for human endeavour. New technologies enable better science and the two bootstrap each other to new heights. But it should always be recognised that science only comes up with theories and not correct or absolute answers. And even if scientific theories are "proven" to be correct at a point in time, they may be subject to improvement or even replacement by better ones that more accurately describe reality or at least our observations and interpretations of reality. This has happened so often in science but not so often in religion even modern religion.

So does it mean that science (theory) is transient while religion (belief) is permanent and if so what does this mean for the future? Will they ever overcome their differences and work cooperatively together or will destructive competition continue to threaten the benefits of either. Can the two branches of human mental endeavour peacefully co-exist and continue to evolve and produce positive outcomes. Perhaps both can continue to be key factors in progressing civilization. But I don't intend to address questions of future competition between science and religion any more. Instead, I believe there is a much more interesting and important topic, "The future for scientific knowledge, theories, laws and constants and the scientific processes behind them". Perhaps it is related to cooperation between religion and science, who knows?

THE FUTURE AND SCIENCE

The future for science is a very broad subject and although very interesting, I don't have the time or skills to even begin to address it. So I want to focus on an even narrower aspect of this overall subject, the future of our current scientific knowledge. In particular how can current scientific work be formalised and then robustly captured and carried forward for us and any others to apply? But is it stable enough or will it evolve too rapidly to be worth capturing in anything other than a transient or volatile way? Will fluidity or change always be the case? Should we even try!

It is well known that human endeavour and scientific knowledge have progressed using collective knowledge. We take what our ancestors gave us and review, revise and improve it using scientific processes. So perhaps of great importance is the question about the validity of current scientific knowledge. Should current theory be captured for all times or should it be left to die out after it has evolved just like in real life? Today most theories are undergoing only minor refinement. Any change seems to be mostly at the periphery. Science seems to have a firm foundation at present with the basic laws seemingly immutable and most theories supporting each other or at least generally being consistent. But it has always been the view of scientists that current theory is correct. However it may just be an illusion like it was so many times before. There may yet be major changes to our basic scientific theories in one or many areas which introduce radical new science. So how do we allow for this dynamic situation and uncertainty?

While addressing the future of our scientific knowledge, the question must be asked, is it really absolute or is it only man made. Is it always "out there" waiting to be discovered just like earth bound humans have been doing or is it just a human creation? Can scientific knowledge readily be derived by any "intelligent life form" just as we have done? Is what we know and continue to discover a priori true in terms of all relative space and time? I am not sure about the time before or during the so called "BIG BANG" but that is another issue. If so then should this absolute knowledge be captured in some absolute way or is it already formalised and readily available but presented in a way we have yet to discover?

Our planet and of course the human species, is relatively insignificant in the infinite universe. We definitely won't last forever, no matter what many optimists say. However, scientific theory and practice developed by humans on earth is, I believe, very valuable. It supposedly applies to the infinite universe and is well worth saving even in its current "work in progress" form. If or when we become extinct, how can life forms that follow us benefit from our efforts of scientific discovery and the development of scientific theory that has resulted? Of course all our mathematical knowledge must be included with our science knowledge. The two fields go hand in hand in a way or is maths just another branch of philosophy?

The collective human knowledge of today is way beyond the capacity of any single human to learn, know and comprehend. But together as a human race we have developed, recorded and continue to apply it to our daily lives. As is the case with today's technology, most can use it freely and easily but very few have any real understanding of how it works. More and more of our knowledge is being concentrated in fewer and fewer brains, leading to greater expertise in more specialised and limited fields. This also poses a threat to our use of knowledge.

Technology has enabled almost all historical ideas and knowledge to be captured and made available in many useful and open forums. It really started with books and has progressed beyond imagination. Large amounts of text/image/graphical dynamic information are now captured and made available quickly, cheaply and globally. In some areas such as military research, information is still kept "secret". In others such as scientific community research establishments, it may not be secret but is often just as inaccessible and hidden away in ivory towers. But most scientific information eventually leaks out and becomes available to everyone.

So what is wrong with the current method of knowledge capture and dissemination? Will it always be readily available to all and is it foolproof and survivable? What is the possibility of corruption, intentional or otherwise? And what about when (not if) a global catastrophe occurs which threatens the future of life on and beyond earth? What if we become the next dinosaur extinction event? Will our only legacy be fossils again?

LIFE (AND SCIENCE) AFTER THE END

Both our planet and all forms of life on it, especially human life, will undoubtedly not last forever. The end of life as we know it may still be way beyond any time we can imagine but it may be sooner that we think or would like. The end of human life here on planet earth may perhaps eventually be largely due to our own misguided efforts. Our explosive population growth has already and will continue to cause massive over consumption of all natural resources. This is resulting in large scale pollution and destruction of our own eco systems which will eventually destroy us if we don't run out of resources first. The scientific development of unnecessary technology (weapons) and bio-chemicals may also cause considerable problems. Weapons of mass destruction in the wrong (human) hands may be catastrophic. But the end may also happen in other natural ways such as a collision with another cosmic body, an unknown solar or even larger super cosmic event. This raises the obvious question of how can we continue our life form, or any life form for that matter, in such a situation? Or perhaps even more fundamental questions are; should we even try to do so, especially if we caused it and will there be sufficient time.

Many scientists have raised the possibility of travel to other planets and even other solar systems to save or extend earth bound life forms. There are even some who have suggested that life on earth may have already come from elsewhere. Then there is the question of what life form would we send out into space, living humans or other life forms? Would they be frozen cells or incomplete life forms or perhaps blue-prints for creating life, if we can produce them?

A major problem is time and energy. It would take very long travel times (with existing technology) to get anywhere. Energy would be a problem if the life forms were to be kept alive. If in blueprint form or some form of suspended animation or frozen form, perhaps energy would not be such a problem in cold space? In any case such ideas are very far-fetched at present but who knows what may happen in another thousand years or so, if we last that long. But what is special about life in general and human life in particular? Knowledge (at least our scientific version of it) seems to have a more important and fundamental basis for continued existence.

According to the current scientific view, science laws and theory are universal (or is it absolute) and the same in all parts of the universe. Who knows, perhaps scientific discovery may have already happened elsewhere and perhaps even much earlier. Some of our science or even more may have already been learned and applied by other life forms. It could have already been sent out for others, such as us, to read or discover. (Just like monoliths "influencing" monkeys in science fiction movies). Would life on earth be any better if we had found such a source of information much earlier on our path toward civilisation or was it necessary that we learned it all from scratch, often at great expense?

Programs like SETI are underway to try and answer these questions. They are looking for extra-terrestrial intelligence (whatever that means) by searching for "unnatural signals" (whatever they are). Perhaps such programs should start looking for natural signals and trying to decode them. But that is where the fun starts because the best coding systems turn signals (almost) into white noise or make it almost perfectly natural. Perhaps we have already been looking at some "good" signals but don't have the correct decoding scheme or language to decipher them.

Perhaps of more immediate interest is how to encode and perhaps send out into space all our scientific (and other?) knowledge so that others, even us, can use it later. The situation is a bit like a time capsule with knowledge being captured and stored in some way so it lasts "forever". We have made and applied many useful scientific discoveries that are effectively larger than life. How can these ideas be continued and carried forward for future life forms to benefit from. Will someone or something somewhere out in the universe one day listen to leaked signals from our "intelligent" life form or recover one of our "time capsules". What would they make of it and what would they do with it? SETI in reverse! Or does everything we have learned about science, and that is supposedly universal, always need to be relearned from scratch by future human or other life forms in another part of the universe. Or would they learn something different?

A significant question is how much do we already know. Do we know everything? Of course not! But what proportion of "total knowledge" do we know. Is scientific knowledge effectively open ended or infinite. The old idea that "most areas of science are already well known so there isn't much more to learn" is long since gone. The current scientific view is that all areas have more questions than can be answered by all the known experts in a lifetime and then some. And of course as experience has shown, for every answer, idea or new discovery, more questions are raised, unfortunately often more complex ones. Perhaps there is more to learn than is supposedly already known. Even worse, perhaps we have gone down a wrong garden path of knowledge with some of our weird scientific theories and we may need to rewind before we can go forward. It wouldn't be the first time this has happened.

THE LIBRARY OF LIFE (SCIENCE)

Most life on earth carries forward information genetically. That is, information on how a life form is to survive and reproduce is retained and propagated within its genetic material (DNA). This in itself is an amazing scientific fact and was a phenomenal scientific discovery. Random changes supposedly occur within DNA as a result of external factors such as radiation. These are supposedly behind the random evolutionary processes. There may also be "improvements to" or "learning" in DNA as result of more internal factors but the details of how this may work are still unclear. Some life forms with higher level brains also teach their young some of life's lessons. Learning mostly via "do as I do" demonstrations are the key to carrying forward many of the processes necessary for higher life forms to survive.

All life forms are "born" with information in their DNA (nature) but it seems that "higher" life forms provide considerable additional information by parental or other teaching (nurture). These two types of information transfer triggered the nature/nurture debate. But is the nurture process and any nurture information captured in DNA and if so then how? The questions became; what is behind some higher life form behaviours and which is more important for survival, nature or nurture? Until recently nurture was considered at least as important as nature for human survival and the basis for many (all?) cultural practices. This is also related to the debate about free will and determinism. How much can we control our own actions by conscious thought versus unconscious automatic (DNA or learned?) control.

The flow and evolution of scientific information from early civilisations, in particular the Greeks, through to eastern cultures and finally on to western cultures has been a fascinating story. Fortunately we have often been able to build on the thinking of others to help develop a sound base of knowledge and theory of almost everything.

We continue to do this today although there are still hiccups in the process. Today with computers, the internet and data libraries like WIKIPEDIA, vast amounts of history and knowledge has been captured. It is constantly updated and is readily available to almost everyone almost instantly. And while most scientific "knowledge" is still in "draft form" or under development and may evolve, the usefulness and value of data on the present knowledge of science is almost immeasurable.

This build-up of knowledge has followed a slow and often tortuous path and its future may be unclear. Can ongoing independent development, accessibility and application of such knowledge, especially scientific based knowledge, be assured for the real benefit of mankind? Where does the knowledge come from and where does it go, who owns it and who controls access to it? Who ensures its integrity and who keeps it in a safe but accessible place? Perhaps of more importance is who decides what is knowledge and what is story (fact or fiction) and how is this applied to the processes of capture, storage and access control. Access to the "truth" has been and will continue to be a crucial factor in how our civilisation evolves but what is the truth? There are many "non-standard" ideas or "non-truths" that have been rejected or are not yet accepted by the mainstream but may also need to be captured in such data bases.

Recently (in terms of human life on earth that is) we have (almost) mastered the ability to capture, apply and more importantly carry forward external knowledge in a more practical and reliable way. First there was language and storytelling often with song and dance teaching methods, then pictures and then the written word which emerged from pictures. This information capture process has evolved, especially in the last few hundred years and has recently greatly accelerated with modern digital processing and storage technology. Now we have become accustomed to and reliant on cheap mass information collection, storage and dissemination systems for much of our information and learning. This information is now being learned directly often without other human (parental) involvement. "Self-learning" is becoming a very important part of the learning process. But are we dehumanising and digitising our nurture process? And is so called self-learning behind the "intelligent robot" threat. And will we soon be capable of digitising our DNA and making designer (GM) babies (Monsters)?

So what is the future of our scientific knowledge and how does it get captured and made available for use by future generations, civilisations and even other life forms, terrestrial or not? I'm not just thinking about next week or next year but more likely next millennium or even greater time scales. Most forms of storage media have a finite life. Even the information sent on Voyager which was encoded in various physical ways from pictures through old gramophone records and other hard copy techniques, won't last very long at all. While the environment of space is relatively sterile and cold and the probability of internal decay or external damage is low, it will be out there for millions if not billions of years before it is "found" if at all. By then it will almost certainly be completely useless in terms of providing any universal knowledge.

How can information be best stored and protected for very long periods? How often should a "snapshot" be taken and frozen in time, just in case? And how should information be encoded so it is future proof and easily available for other intelligent life forms to access? Undoubtedly new digital technologies and software systems play a major role in this regard. Systems available today are capable of safely and securely storing almost all known information and making it readily available in future, or so we think. But these systems are still evolving and todays systems will be primitive in even a few years. What is the integrity of data stored on old technology developed for use in short term commercial and informal environments?

WHERE IS VOYAGER

The question of sending information out into space has already been considered. One of the most fascinating events in my lifetime was the launch of the Voyager spacecraft. They were sent off into space just after the Apollo lunar missions were undertaken. By now these spacecraft have just about left our solar system and will eventually drift off into deep space. This alone is an amazing achievement. Communication with them is still possible but not for much longer due to energy and distance problems. The already faint signals will eventually get too weak to receive over noise and the nuclear power will eventually run out. While they may be travelling at high speed relative to our solar system, they are moving slowly in the vastness of space. Who knows if and when they may eventually pass another body in our galaxy or beyond and perhaps be captured? And if one does end up in some intelligent life forms backyard how will they make sense of it. All very speculative and extremely unlikely but it's good to dream.

While I was excited at the time and continue to be amazed at many aspects of the Voyager saga, what made me cringe was the information that was included. I know it was only a last minute idea and it was a bit like a school yard time capsule exercise, but it was not very scientific, even at that time. It raises questions about the types of information that should be selected for any future space/time capsule venture. Any spacecraft launched into deep space can carry data but for what purpose and who will use it? Where is it going, how long will it take to go anywhere realistic and will it even last the distance? They may even collide with other space debris, not man made this time, and be destroyed

Even at the most optimistic estimate, it will take billions of years to get anywhere. By then having drifted in deep space and been bombarded by cosmic energy and particles of various types what will be left to decipher. Perhaps nothing will last and everything will be broken up by "old age", disintegrate to dust and just fade away. This is likely to happen well before they even approach any other major cosmic body or go anywhere near any other solar system. Even if they did, the most likely outcome would be capture and destruction by another sun. The information (if that is a meaningful term) will almost certainly be lost in the noise. So in reality the truth is likely to be very different to the original intention. The "information" contained in the spacecraft will be lost and all the good intentions will come to nothing. Was it worth it? Yes, I believe.

The question of where we send our information is perhaps less important than the questions of what we send and how we send it. What types of information should be captured and how should it be presented. Information about human and other life forms, scientific theories and laws and any other "knowledge" on us, our planet and our solar system can be "captured" and "presented" in many ways. Some well-meaning but "weird" information was included in the Voyager "time capsule". Most of it was hard copy physical information. Perhaps the most un-scientific information was a plastic circular disc (LP record) which captured pop music in simple analog grooves. There was no real attempt to capture bold scientific information in a soft copy way using digital technology, although that was not very advanced at that time. What would we make of it in a few thousand years, if we still exist then? What would other life forms make of it if captured billions of years from now at some far reaches of our galaxy or beyond?

If we were to send out another type of Voyager time capsule what would we send today? Would it only be digital information? Should we decide against sending out physical time/space capsules and use virtual ones instead? A soft copy package of information in a telecommunication burst could be sent out like a digital message in a digital (quantum) light "bottle"? A least light (telecommunication) travels a bit faster than a physical spacecraft and can be sent in many directions at once and easily repeated. But what would be included and how would it be encoded and transmitted?

Technology was relatively primitive in Voyager times compared with today's technology. And today's technology will be primitive compared to future technology, especially in another 1000 years. How can we prepare for the future or should we wait? Our knowledge is changing as fast as technology. It is a bit like the current process of backing up data in a soft copy (computer) environment. Backup is done regularly so as to capture even incomplete data and incremental changes in information to protect from catastrophic failure. This is big business for telecommunications, data management and computer companies. Should we transmit (backup) what we have now or wait until all the data or theories are complete?

So in general terms how can data or information can be carried forward for future epochs. How should it be encoded, stored, protected and if necessary encrypted before it is stored locally and also sent out into space. How long will it last: thousands of years, seems likely; millions of years, unlikely; billions of years impossible unless .....?? These are some of the questions science could be addressing now. Perhaps it should be coupled with the question of where human and other life on earth will be in a million or so years. Our solar system, let alone our planet, won't be here for ever. It has a very finite life or at least a finite period of human habitation. Where are we planning to go?

SCIENCE BEYOND

What does the future really hold for science? Will mankind survive for the next thousand years and beyond? Will science outlive us? Perhaps neither, hopefully both. But is our science knowledge absolute or just a fragment or approximation (Greek shadow) of the real truth?

Can science be simplified and are there a few basic theories and models with fewer assumptions and "natural" constants or is science open and unbounded? Perhaps the focus will move to the search for the so called "end game" such as Grand Unifying Theories. If so will "free will" science suffer? Will science become more driven by commercialism, short term gain and direct financial benefit or will "blue sky" research survive and thrive. And will science come under the dark of bigotry like so many times before or be illuminated by a bright future? Illumination by rose coloured or any other distorting light must be avoided but perhaps there will always be some illusions in our efforts. There are examples of misguided effort caused by over enthusiastic scientists, but does misguided effort sometimes progress science?

And what happens when some new ideas like mine come along. Will they be considered to have merit and even become "accepted"? What will they mean for physics in general and standard models of the atom and the cosmos in particular? Will things suddenly change in any physical sense? Well the answer to the last question is hopefully no. The world we live in shouldn't change at least not in a major way, even if we find a better way to describe or model it. And what is the benefit of scientific research, especially so called theoretical research? How far should research and development proceed in certain sensitive human areas like human DNA experimentation and human cloning? Should they be allowed and even encouraged?

Perhaps discussion on these so called taboo topics will initially be purely philosophical, a bit like early Greek science. It may be less rigorous at first but in due course as the questions and outcomes are more formally defined, may become a more serious branch of science. The ongoing debate about the real benefit of such science and the real dangers or threats of new scientific ideas will be very important.

Whatever the future holds, the work of science is so important today that it must continue. It must be protected against "erosion" by bad influences and be encouraged (financed) by state and society. And it must be open to new ideas, yes even maverick ones like mine. There are examples of over enthusiastic scientists chasing dreams. But science must avoid cults or becoming like a "religion", just as it must avoid the creeping inroads of commercialism and "justification". These factors may distort the path of pure science in a damaging way. Science should be made as open as possible for all to participate and for all to benefit from. Hopefully everyone will do more on both counts? What does science mean to you and how will you participate and benefit? In any case, all the best for your science based future.

But before you put this book down, here is a word about my next book. This time it is on a completely unrelated subject. I will leave physics for a while and examine a completely different subject, human civilisation. In particular I will examine the evolution of many "isms" including local and global commercialism and in particular how the allocation of work and distribution of wealth occur. I want to find causes of and solutions to the many inequalities of life and especially try to understand the concept of richness and poverty if that is possible. Such inequalities have been and will continue to be the main cause of disturbance in an otherwise peaceful existence. The aim will be to try and determine how to better manage the future we are making for ourselves and our offspring and improve our existence. Big bold broad ideals but they are always the best ones. I haven't finished or even started it yet so you may have to wait a short while but again the wait will be worth it.

Printed in the United States
By Bookmasters